'Sustainability' may be the most important word i misused This brilliant, deep, accurate, well-refere rectify that. It should be required reading for ever politician.

Paul and Anne Ehrlich, *Stanford University, USA*

Demystifying Sustainability bids the modern world to abandon magical thinking. It provides just what the world needs most in its time of gathering crisis – a viable, alternative way-of-being that both celebrates the full spectrum of human potential and accurately represents the biophysical *realities* within which the human animal must function.

William Rees, *University of British Columbia, Canada*

Haydn Washington brings wide reading and experience, plus a sharp mind and sharp knife, to cut away the obfuscation and denial that mystify the discussion of sustainability. The book is accessible without any 'dumbing-down', and does not evade the difficult questions of overpopulation, overconsumption, deification of growth, and overconfidence in the ability of technology to substitute for morality. Highly recommended.

Herman Daly, *University of Maryland, USA*

Washington boils down the 300 definitions of sustainability to distill the potent essence of the term, and then applies it to point us in the direction of a better future.

Erik Assadourian, *Worldwatch Institute, USA*

We want a future that is not only sustainable, but also desirable – a future that allows a prosperous and equitable economy embedded in a harmonious society that remains within planetary ecological boundaries. Our current 'growth at all costs' system and its trajectory are neither sustainable nor desirable. Haydn Washington has provided a powerful demystification of sustainability and a guide to the future we really want.

Robert Costanza, *Australian National University, Australia*

It is hard to have a serious discussion about a subject as important as sustainability when we don't even agree on what it means. Harder still if we don't even know that we don't agree on its meaning. Haydn Washington understands this difficulty only too well and his book succeeds grandly in demystifying sustainability.

Peter A. Victor, *York University, Canada*

This book does not beat around the bush. It does not talk in circles. It does not pretend that sustainability is so difficult to grasp that it cannot be achieved. Nor does it pretend that it can be achieved by following business-as-usual trajectory. It does name problems by their name. It does ask difficult but necessary questions. It does propose real solutions. This book is a must read for everybody, in both so-called developed and developing countries, from students to policy-makers.

Helen Kopnina, *The Hague University of Applied Science, The Netherlands*

'Sustainable' must not remain merely an undefined buzzword of our time. The human predicament requires widespread understanding that humanity is *in serious trouble* because we have overshot this finite planet's *sustainable* human carrying capacity. Only by immediately outgrowing the 'endless growth' myth can we avert a fatally calamitous future for our species.

William Catton, *Washington State University, USA*

Sustainability and spirituality are the double helix of our future wellbeing. Rekindling a sense of wonder in our natural environment is essential for the quality of life of our increasingly urbanized society. This book lives up to the promise of its title and goes a long way towards demystifying sustainability for a broad audience.

Julian Crawford, *International Society of Sustainability Professionals, USA*

Haydn Washington has nicely deconstructed the dominant economic paradigm that is driving us toward global collapse. Alternatives are clearly explained through a tool kit of out-of-the-box thinking that could save us from ourselves. I recommend it to all who care about a quality future for life on Earth.

Colin L. Soskolne, *University of Alberta, Canada*

DEMYSTIFYING SUSTAINABILITY

What *is* sustainability? Much has been said about the terms 'sustainability' and 'sustainable development' over the last few decades, but they have become buried under academic jargon. This book is one of the first that aims to demystify sustainability so that the layperson can understand the key issues, questions and values involved.

Accessible and engaging, *Demystifying Sustainability* examines the 'old' sustainability of the past and looks to the future, considering how economic, ecological and social sustainability *should be* defined if we are to solve the entwined environmental, economic and social crises. It considers if meaningful sustainability is the same as a 'sustainable development' based on endless growth, examining the difficult but central issues of overpopulation and overconsumption that drive *un*sustainability. The book also explores the central role played by society's worldview and ethics, along with humanity's most dangerous characteristic – *denial*. Finally, it looks to the future, discussing the 'appropriate' technology needed for sustainability, and suggesting nine key solutions.

This book provides a much-needed comprehensive discussion of what sustainability means for students, policy makers and all those interested in a sustainable future.

Haydn Washington is an environmental scientist and writer of 40 years' experience. He is a Visiting Fellow at the Institute of Environmental Studies at the University of New South Wales, Australia.

DEMYSTIFYING SUSTAINABILITY

Towards real solutions

Haydn Washington

First published 2015
by Routledge
2 Park Square, Milton Park, Abingdon, Oxon OX14 4RN

and by Routledge
711 Third Avenue, New York, NY 10017

Routledge is an imprint of the Taylor & Francis Group, an informa business

© 2015 Haydn Washington

The right of Haydn Washington to be identified as author of this work has been asserted by him in accordance with sections 77 and 78 of the Copyright, Designs and Patents Act 1988.

All rights reserved. No part of this book may be reprinted or reproduced or utilised in any form or by any electronic, mechanical, or other means, now known or hereafter invented, including photocopying and recording, or in any information storage or retrieval system, without permission in writing from the publishers.

Note: The cover photo of the book is by the author, taken at a pool in the Colo Wilderness in Wollemi National Park, NSW, Australia.

Trademark notice: Product or corporate names may be trademarks or registered trademarks, and are used only for identification and explanation without intent to infringe.

British Library Cataloguing-in-Publication Data
A catalogue record for this book is available from the British Library

Library of Congress Cataloging-in-Publication Data
A catalog record for this book has been requested

ISBN: 978-1-138-81268-0 (hbk)
ISBN: 978-1-138-81269-7 (pbk)
ISBN: 978-1-315-74864-1 (ebk)

Typeset in Bembo
by Wearset Ltd, Boldon, Tyne and Wear

For the 'Js' – Jessie, Joyce and Jeannie

CONTENTS

List of figures xiii
Foreword xiv
W‍illiam R‍ees
Acknowledgments xx

Introduction: sustainability – seeking clarity in the mist 1

1 The 'old' sustainability: a story of listening and harmony 6
 A history of caring 8
 Modernism and the decline of harmony 11
 The revolt against modernism 12
 Conclusion 14

2 The 1960s to the present: key conferences and statements 17
 The fabulous 1960s and 1970s 17
 Key conferences and reports 20

3 Rise of the 'new' sustainability – the weak and the strong 29
 Destination or journey? 29
 Is 'sustainable development' a code for sustainable growth? 30
 Does 'sustainable development' equal sustainability? 33
 Sustainability – the 'weak', the 'strong' and the 'strongest' 38
 Conclusion 42

4 Economic sustainability: coming to grips with endless growth 47
   The underlying assumptions of neoclassical economics 51
   The steady state economy 54
   Degrowth and the circular economy 55
   Endless growth and 'growthmania' 56
   Reductionism, economic modelling and the 'fallacy of misplaced concreteness' 58
   'Resources are infinite', techno-centrism and substitution 59
   The ethics of economics 60
   The 'green' economy 61
   What should economic sustainability mean? 62

5 Ecological sustainability – essential but overlooked 71
   Do we have a problem? 72
   Human dependence on Nature 73
   Theory and the 'balance of Nature' 79
   What should ecological sustainability mean? 86

6 Social sustainability – utopian dream or practical path to change? 94
   Introduction 94
   Utopia 95
   What is fair? Equity and equality 96
   What is just – and for whom? 99
   Social cohesion and capital – keeping it 'together' to act 100
   Democracy 101
   Governance 101
   War and conflict 104
   The practicality of social sustainability 104
   What should social sustainability be? 106

7 Overpopulation and overconsumption 114
   Overpopulation 114
   Dealing with consumerism and overconsumption 119
   Dematerialising our economy 122
   Beyond 'triple bottom line', 'eco-efficiency' and the 'small and easy' approach 125
   Alternatives to the consumer society 129
   Dealing with the heresy of more 129

8 Worldview and ethics in 'sustainability'   136
   Worldview, ethics, values and ideologies 137
   Anthropocentrism 141
   Anthropocentrism in 'sustainable development' 143
   Ideologies – modernism 144
   Intrinsic value and the revolt against modernism 144
   Deep ecology 145
   Ideologies – postmodernism 146
   Questioning reality and 'Nature scepticism' 147
   A sense of wonder 149
   Anti-spirituality in Western culture 151
   How do we bridge the great divide? 151

9 An unsustainable denial   159
   Believing in stupid things 160
   Scepticism vs denial 161
   Denial is common 161
   The history of denial 163
   Ideological basis for denial 166
   Psychological types of denial 167
   Ways we let denial prosper 170
   Conclusion 172

10 Appropriate technology for sustainability   177
    Appropriate technology 177
    Renewable energy technologies 180
    Inappropriate technologies 184
    Conclusion 187

11 Solutions for sustainability   191
    Can we know what future generations will want? 191
    Solutions – pluralism vs specificity 192
    Demystification – what meaningful sustainability cannot be 193
    What sustainability should be? 195
    Is it too late? Optimism, pessimism and realism 196
    Solutions 199
    1 Worldview, ethics, values and ideologies 199
    2 Redesigning ourselves to enable change 200
    3 Population 201
    4 Consumerism and the growth economy 202
    5 Solving climate change 202
    6 Appropriate technology: a renewable future 203

7 Reducing poverty and inequality 204
8 Education and communication 205
9 The politics of it all! 207
The 'Great Work' 208
Can we demystify 'sustainability'? 208

Index 215

FIGURES

3.1	The 'sustainability trinity'	30
4.1	Nature, society and economy	48
4.2	The assumed 'circular flow' of production and consumption in the neoclassical economy	53
5.1	Simple food web for the Australian platypus	74
5.2	The global nitrogen cycle	77
5.3	The relationships between ecosystem health and human well-being	78
6.1	Index of health and social problems vs income inequality in world nations	98

FOREWORD

William Rees

UNIVERSITY OF BRITISH COLUMBIA, ORIGINATOR AND CO-DEVELOPER OF THE 'ECOLOGICAL FOOTPRINT' CONCEPT

On 'getting real'

This book is about sustainability and how to achieve it. But it is also about two other seemingly unrelated things – the 'nature of reality' and 'humanity as work-in-progress'. The facts are that humans have a limited grasp of reality. Worse, 'We have all by our actions or lack of them – in particular over the last [almost half-century] – agreed to deny reality' (Saul 1995, p. 18). Fortunately, our species is still evolving. *Homo sapiens* is unfinished and improvable – but still subject to the demands of natural selection. Both perceived reality and human evolutionary status bear on prospects for sustainability. (You should be worried!)

Three indisputable facts frame the analysis. First, since the dawn of agriculture, humanity's material and cognitive relationship with the rest of the natural world has become increasingly dysfunctional. The accumulating scientific evidence shows that humanity is now on a collision course with biophysical reality. Second, *Homo sapiens* has the high intelligence, analytic ability, organisational skills and material resources needed to avoid catastrophe; third, despite such abilities, the global community has failed utterly to take the necessary evasive action. On the contrary, world leaders, in apparent denial of the evidence, regularly unite in a vigorous chorus of 'steady as she goes'.

Demystifying Sustainability goes a long way toward explaining this conundrum and its origins in unique dimensions of human nature. 'Perception' is at the heart of the problem. Humans rarely perceive anything just as it is. From birth, every sensory input and social encounter a person experiences contributes to the formation of various cognitive filters through which that person interprets all subsequent experiences. Thus, two individuals with different family, social, or educational backgrounds are likely to interpret subsequent identical inputs in disparate ways. In effect, individuals and social groups create their own 'realities'. Consider how many differing cultural narratives, tribal myths, religious doctrines, political ideologies, disciplinary paradigms, ethical frameworks and other worldviews exist to divide the human family. We humans are, by nature, story-tellers and myth-makers who collectively, selectively and subconsciously make things up as we go along.

Sociologists refer to this process as the 'social construction of reality' (Berger and Luckmann 1966)[1] but it can function equally as the 'social construction of mass delusion'. Each 'social construct' is a product of mind first birthed in words and subsequently 'massaged or polished by social discourse and elevated to the status of *received wisdom* by agreement among members of the social group who are creating the construct' (Rees 2013).[2] It is important to note that while 'what we know' may masquerade as reality: (1) *all* formal knowledge is to some degree socially constructed; (2) not all versions of 'truth' can be equally valid; and (3) many objects and phenomena (e.g. the laws of gravity, motion, and thermodynamics) *really* exist regardless of whether or how people conceive of them.

Not surprisingly, there are myriad conflicting definitions of 'sustainability' and 'sustainable economic development' (note the three layers of social construction). Some analysts, including Haydn Washington, see *Homo sapiens* as an evolved species that remains an integral, dependant part of nature and whose societies and economies must be structured to reflect these basic biophysical facts. To them, sustainability requires that human societies learn to exist in harmony with the ecosystems that sustain them. Or perhaps 're-learn' – Dr Washington points out that the feeling 'of being part of Nature, of believing Nature is sacred ... [of needing to maintain] the harmony of the Universe, has emerged again and again in cultures across the world'. Ancient peoples learned, often the hard way, to think of the human–nature bond in terms of harmony, balance, reverence, sacredness, respect, custodianship, stewardship, beauty, and even *love*. A modern economy based on such values would necessarily mimic the 'steady-state' material cycles and energy flows through nature; humans would neither take more than ecosystems can produce nor discharge more wastes than they can assimilate on a continuous (i.e. 'sustainable') basis. An overriding principle would be the maintenance of biophysical life-support functions. It is hard to imagine an economy so conceived *not* being sustainable.

At the other end of the belief spectrum are those who argue that humans have transcended their lowly biological origins. By this construct, people are no longer part of, or dependent on nature, and owe no allegiance to the natural world. Such human 'exceptionalism' dissolves fear of climate change, the ecological crisis and resource shortages. Indeed, human ingenuity and technological prowess are the only resources that matter, infinitely capable of seeing us through any temporary set-back. Listen to the ebullient confidence of the late Professor Julian Simon:

> Technology exists now to produce in virtually inexhaustible quantities just about all the products made by nature.... We have in our hands now – actually, in our libraries – the technology to feed, clothe, and supply energy to an ever-growing population for the next 7 billion years.
>
> *(Simon 1995)*

Admittedly extreme (and arithmetically absurd), such assertions nevertheless feed the widespread contemporary belief that humanity has freed itself from material bonds to the living Earth (emotional bonds never enter the discussion). Thus liberated, the global mainstream has fabricated a 'sanitised' (Washington's word) version of sustainability that shifts the focus from harmony, integrity and caring to the cold mechanics of perpetual growth monitored by GDP per capita with occasional reference to 'carbon footprints', 'energy audits' and related forms of routine bookkeeping.

It doesn't end there. The operational platform for mainstream sustainability is corporate capitalism and its hand-maiden, neoliberal market economics. The neoliberal construct gives us a lifeless, mechanistic model of the economy concerned mainly with maximising the efficiency and productivity of a 'self-generating circular flow of exchange value' (the flows of money back and forth between firms and households) and the accumulation of human-made capital. With remarkable conceptual sleight of hand, economists have thus constructed an abstraction of the economy that floats free from both the ecosystems within which the economy is embedded *and* the human community it purports to serve. Moreover, seeing the economy as a circular money flow with no attention to unidirectional energy and material throughput is akin to considering the human body as a circulatory system with no reference to the digestive track. One might as well ask an engineering student to accept that 'a car can run on its own exhaust' (Daly 1991, p. 197).

As telling, the ethical stance of neoliberal economics toward the living world is '*utilitarian, anthropocentric* and *instrumentalist*.... It is utilitarian in that things count to the extent that people want them; anthropocentric, in that humans are assigning the values; and instrumental, in that biota is regarded as an instrument for human satisfaction' (Randall 1988, pp. 217–218, original emphasis). As final insult to reality, the model assumes that market outcomes are determined by rational individuals expressing fixed consumption preferences, where 'rational' is defined as maximising personal 'utility'. The model thus reduces people to atomistic consuming machines devoid of family, community, sense of place and empathy toward the natural world. Ironically, the worldview that breathes life into mainstream economics sounds the death knell for ecological integrity. Dr Washington remarks that the only connection between mainstream economic theory and the living world is that perpetual growth remains the largest elephant in the room.

It should be easy for reasonable people to choose between competing visions of sustainability – but there is a complication. So-called 'post-modernists' claim that, since *all* knowledge is socially constructed there can be no objective reality; modern science, therefore, has no greater claim to truth than alternative ways of knowing. Indeed, extreme relativism would assert that lunacy and science have an equivalent grasp on reality.

This is dangerous psycho-babble. It is particularly nonsensical when the competing constructs pertain to concrete entities and real processes that exist apart from human imaginings. To be useful in addressing problems in the real world, 'words and concepts must be anchored in external reality'. Indeed, 'it is the key to intelligence if not sanity to be able to assess with some accuracy the extent to which words [doctrines and myths, ideologies and paradigms] refer to the world of non-words' (Postman 2000, p. 75).[3] As for social construction, 'You may say, if you wish, that all reality is a social construction, but cannot deny that some constructions are truer than others' (Postman 2000, p. 76).[4]

All of which is to say that the world community must surely stand with Haydn Washington in contemplating the prospects for sustainability on this finite planet. His is a call to 'get real'; the vision he projects in *Demystifying Sustainability* is the 'truer' vision. *Homo sapiens* really did evolve in, and remains a dependent component of, real ecosystems. Untainted by waffling relativism, the implications of these simple truths are game-changing. Even as economists' monetary abstractions imply that humanity is decoupling from nature, real-world material flows show that *Homo sapiens* is now the most ecologically significant herbivore *and* carnivore in all the world's major ecosystems. Our natural expansionist tendencies, reinforced by the twin cultural myths of infinite technological progress and perpetual economic growth, have made of us the ultimate consumer species. *Homo sapiens* has become dangerously parasitic on the Earth, the combined demands of bio-metabolism and industrial metabolism literally consuming the ecosphere from the inside out and undermining basic life-support. Any vision for a future sustainable society that does not start from these biophysical realities is not of this world.

Demystifying Sustainability bids the modern world to abandon magical thinking. At face value, this does not seem a radical proposition but may prove difficult in the execution. Cognitive neuroscience reveals that the shackles of socially constructed concepts can be difficult to shake off. Repeated exposure to culture-specific beliefs creates corresponding synaptic circuits in the developing brain. (Ironically, even socially constructed illusions acquire a *real* physical presence.) Subsequently, people tend to seek out compatible experiences and, 'when faced with information that does not agree with their [preformed] internal structures, they deny, discredit, reinterpret or forget that information' (Wexler 2006, p. 180). For a rapid shift in perceptions, engrained illusions may have to be shattered by some catastrophic event. The slower alternative – voluntarily freeing oneself from deeply entrenched but erroneous beliefs – requires that the individual acknowledges the error of his/her ways and works hard to construct a more adaptive worldview.

While synaptic imprinting alone is enough to inhibit the needed paradigm shift, there are other significant psychological barriers. For example, humans are naturally myopic – we favour the here and now over the future and distant places – a trait that economists have formalised in the concept of 'social discounting'; when safety or 'survival' (including self-esteem and socio-economic status) are threatened, instincts and emotions that operate *beneath consciousness* may override reasoned responses. (The rich and powerful readily sacrifice the long-term common good in defense of their elitist status.) So it is that at this stage in our evolution, *Homo sapiens* is a deeply conflicted species – we think we act mostly from cerebral consciousness but can count on high intelligence to prevail only in situations free of emotion or social programming. If rapid cultural change proves necessary, our psychological investment in the status quo may prove a formidable stumbling-block. Only climate catastrophe or some other massive ecological shock will be enough to discredit today's expansionary cultural narrative and force the world community to hit the re-set button.[5] Will it then be too late?

More optimistically, if Mother Nature is extraordinarily generous, she may yet give science and protest politics enough time to nudge society toward an orderly transition. After all, humans remain uniquely capable of logical thought, of reasoning from the evidence and of using the results to plan ahead; and no other species comes close to *H. sapiens*' array of flexible mechanisms for cooperation in achieving common purpose. It helps that millions of people in the streets can be politically persuasive. What more does the world need to shed its fantasies, come to its senses and evolve a more viable global culture?

Which brings us back to *Demystifying Sustainability*. This book describes an alternative way-of-seeing that both celebrates the full spectrum of human potential and clearly represents the biophysical *realities* within which the human animal must function. People being pushed in shock, confusion and dismay from the collapse of conventional wisdom also need the pull of an attractive option to help catalyse the transition. Haydn Washington's version of sustainability provides such a magnet, replete with the economic security, ecological stability and social cohesion increasingly lacking from techno-industrial society.

In the final analysis, nothing stands in the way of sustainability but human nature – behavioural traits that were once robust have become maladaptive in the very environment they have helped to create. If, in the crunch, primitive emotions, tribal instincts and culturally-programmed denial succeed in their defence of today's ecologically naive expansionist world-view, then nature's great unfinished experiment with high intelligence may well be shut down. We know from the implosion of previous civilisations that natural selection weeds out defective memes (units of cultural information) just as effectively as it does maladaptive genes (units of biological information). In short, if modern global society fails to 'demystify sustainability' and act accordingly, it may be just as unceremoniously 'selected out' as were the Sumerians, the Mayans and many societies in between.

Notes

1 This is actually a misnomer. Humans socially construct their perceptions, which may or may not comprise an accurate representation of corresponding reality.
2 The propensity to create shared 'realities' (or illusions) is a highly adaptive trait. Social norms are necessary for civil society; common visions, myths and narratives contribute to group identity, social cohesion and individual security within the tribe.
3 If one really believed that gravity was a mere social construct, then s/he should have no fear of leaping from a tall building.
4 Philosopher Karl Popper put the point this way: 'What the scientist's and the lunatic's theories have in common is that both belong to conjectural knowledge. But some conjectures are much better than others' (Popper 1972).
5 As of September 2014, monster storms, record floods and acid oceans have yet to impress; time's arrow has witnessed only a series of failed climate negotiations.

References

Berger P. L. and Luckmann, T. (1966) *The Social Construction of Reality*, Garden City, NY: Doubleday.
Daly, H. E. (1991) *Steady-State Economics* (second edition), Washington, DC: Island Press.
Popper, K. (1972) 'Conjectural knowledge: my solution of the problem of induction', in *Objective Knowledge: an Evolutionary Approach,* ed. K. Popper, Oxford: Oxford University Press.
Postman, N. (2000) *Building a Bridge to the 18th Century*, New York: Alfred A. Knopf.
Randall, A. (1988) 'What mainstream economists have to say about the value of biodiversity', in *Biodiversity*, ed. E. O. Wilson, Washington, DC: National Academy Press, pp. 217–223.
Rees, W. E. (2013) 'Confronting collapse – human cognition and the challenge for economics', in *Confronting Ecological and Economic Collapse*, eds. L. Westra, P. Taylor and A. Nichelot, London and New York: Routledge and Earthscan.
Saul, J. R. (1995) *The Unconscious Civilization,* Concord, Ont: House of Anansi Press.
Simon, J. (1995) 'The state of humanity: steadily improving', *Cato Policy Report*, Washington: Cato Institute, online, available at: www.libertarianism.org/publications/essays/state-humanity-steadily-improving.
Wexler, B. (2006) *Brain and Culture*, Cambridge, MA: MIT Press.

ACKNOWLEDGEMENTS

I would like to thank Routledge (Earthscan) for its support for this book, in particular Khanam Virjee, Commissioning Editor – Development Studies and Sustainability, and Bethany Wright, Editorial Assistant. I would like to acknowledge all those who assisted me to develop the broad scope of this book. In particular I would like to thank one of my reviewers, Dr Helen Kopnina, Coordinator of the Sustainable Business programme at the Hague University of Applied Science, Netherlands, who was particularly helpful with comments and suggestions. I would also like to thank Associate Professor Mark Diesendorf of the Institute of Environmental Studies, for both his comments on Chapter 10 on appropriate technology, and for his discussion of sustainability issues in general. I would also like thank my other reviewers (who remained incognito) for useful comments and insights. I would like to thank Professor Stuart Hill of the University of Western Sydney for comments and suggestions in regard to 'redesigning ourselves' in Chapter 11. I would like to thank Professor Herman Daly and Dr Brian Czech of the Center for the Advancement of a Steady State Economy for discussions about the steady state economy in Chapter 4. I would like to thank Dr Brian Miller of the Denver Zoological Foundation for discussions regarding ecological theory in Chapter 5. I would like to thank Mr John Seed for discussions about deep ecology and environmental ethics over many years. I would also like to thank Dr David Roser of the UNSW Water Research Centre for discussions over some years about the nature of 'sustainability'. Regarding the figures used in the book, I would like to thank Professor Peter Victor for permission to adapt Figure 4.1 from his book 'Managing Without Growth'. I would like to thank Professor Herman Daly for permission to adapt Figure 4.2, from his book 'Steady State Economics'. I would like to thank Mr Max Oulton of the Waikato University in New Zealand for the use of the nitrogen cycle diagram in Figure 5.2. I would also like to thank the Millennium Ecosystem Assessment for permission to use Figure 5.3, and Professor Richard Wilkinson for permission to use data taken from the Equality Trust for Figure 6.1.

INTRODUCTION

Sustainability – seeking clarity in the mist

> *We live today in an age of sustainababble, a cacophonous profusion of uses of the word sustainable to mean anything from environmentally better to cool.*
>
> *(Robert Engelman 2013, p. 3)*

It is time to demystify *sustainability*, to seek clarity. Is 'sustainability' like the weather, where everybody talks about it, but nobody does anything? Almost everyone uses the word, but is its meaning clear? Is it the key term that will lead us to a workable and liveable future? Is it the single most important concept to emerge (or re-emerge) in Western society in recent decades? Alternately, have we made a radical mistake as to what sustainability *is* (Foster 2008, p. xiii)? Has it become a meaningless buzz word? Or even worse, has it become 'sustain*ababble*' (Engelman 2013)? Has the term been co-opted by those who seek to use it for their own purposes, and turn it into mere tokenism, while they continue business-as-usual? Has it become a plaything of academic discourse (and ideologies) around which academics then develop arcane jargon the layperson cannot understand? As one of the authors of UNEP's 'The Economics of Ecosystems and Biodiversity' has noted, sustainable development is often 'more talk than action' (Sukhdev 2010). Is it all of these? This book will seek to demystify the topic, delve into these meanings, and examine what sustainability is (and is not), and what it *should be*. It is hard to think of a more crucial topic in today's world, a world with an environmental crisis, an ecological footprint of 1.5 Earths (GFN 2013) and a species extinction rate at least 1,000 times above the norm (MEA 2005).

So what do we *mean* by 'sustainability'? There are well over 300 definitions of sustainability (Harris and Throsby 1998), and indeed MacNeill (2006), who worked on 'Our Common Future' (WCED 1987), suggested that a new way to define 'infinity' was the ever expanding number of interpretations of 'sustainable development'. One should not assume that people in fact mean the same thing when they use the word. The terms 'sustainable development' and 'sustainability' have been said to reflect the growing groundswell of concern about what is happening to the world (Soskolne et al. 2008) – an environmental crisis of unparalleled magnitude. Even in 1969, the US National Environmental Policy Act said it was government policy to 'create and maintain conditions under which man and nature can exist in productive harmony' (Engelman 2013, p. 7). Hence one description of sustainability is that it is 'an attempt to provide the best outcomes for the human and natural environments both now and into the indefinite future' (EPA 2013). One can of course debate what 'best' means. 'Best' might be said to be about maintaining the diversity and creativity in both human culture and Nature. At the heart of the concept of sustainability should be a vision of achieving human and ecosystem well-being *together* (MEA 2005). Of course one can also debate what 'wellbeing' is, and the simple answer to this would be that this is human culture and Nature both continuing as vibrant, diverse and creative systems. We should aim for our culture to be like this, and we should also seek for Nature to continue that way. Why? Because Nature runs the ecosystem services that support our society, but also because it enriches our lives, ethics and culture. Another ethical reason we should want this is because we love and respect Nature and believe in its 'intrinsic value', its right to exist for itself.

'Sustainability' is a fascinating topic full of contradictions, tangled meanings, unstated assumptions and confusion. Indeed it seems that many people are at cross-purposes and use the word 'sustainability' to push their *own* particular agendas. There is also a frustrating 'looseness' in terminology. We will discuss whether 'sustainability' and 'sustainable development' are actually the same thing. This is either assumed to be true, or subject to a debate involving a surprising passion. Some think so, others argue sustainability is the goal and sustainable development the path. Others think they are related but separate. Others claim that 'sustainable development' is an *oxymoron*, a contradiction in terms, that you cannot continue to 'develop' and yet be sustainable, since we have degraded ecosystems in so many major ways. Others again claim in response that it depends what you mean by 'development', that it can be 'qualitative' development that does not use resources, not quantitative growth. Given that the word is used by many different disciplines, we will also need to ask 'sustainability for whom?'. Or for what? What exactly is it that we are keeping *sustainable*? The lifestyle we value? Our society? The nation's GDP? Or is it the ecosystems and the ecosystem services on which society depends? Or is it Nature as a whole, of which we are a part, and to which we owe respect and care?

There has been a trend within academia and society to keep the term 'delightfully vague', as if that vagueness makes 'sustainability' more adaptable. As if by not defining the term, we might get society to at least act to solve *some* parts of the environmental crisis, without really considering the root causes, so that we might get some action without challenging the 'endless growth' worldview that drives most environmental problems. Some have claimed this was a 'clever strategy', but one can also wonder if in so doing it has not avoided the hard issues? Thus 'sustainability' can come close to meaning 'all things to all people', so it can be argued it has come to mean nothing. Any concept that has to encompass almost everything must lose its own meaning (Ott and Doring 2008). Part of the problem is that commonly people do not define what they *mean* by 'sustainability' or its component parts. Partly it is that they don't start by considering where they come from philosophically and ethically. What are the worldviews, ethics and values that underpin what they mean by the term? Thus the literature is full of people talking about a 'sustainability' that is their own (undefined) meaning of the term, while others speak of *their* conception of the term. No wonder people get confused! In this book I shall seek to cut through the mist around 'sustainability' and focus on key solutions. I shall move towards meaningfully defining 'sustainability' and its component parts: ecological, economic and social sustainability. Furthermore, I'll offer an ethical judgement as to what they *should* be if our human civilisation is to continue on Earth into the long-term.

To return to sustainability's meaning, it is this vision of *joint* human and ecosystem well-being that I argue is essential, and the lack of which has led to the environmental crisis. Re-establishing this vision (as the core of sustainability) I believe is a crucial part of solving the environmental crisis. We shall see however that others don't think this is what sustainability is about. This is why this is a conversation we must have. This is an essential issue, an issue I invite the reader to themselves demystify, and so take real and effective action. We cannot meaningfully discuss sustainability unless we discuss population and consumption, as we do in Chapter 7. We need also to talk about 'worldview' and ethics if we are to talk meaningfully about sustainability, as we do in Chapter 8, and about a very human failing: *denial*, as we do in Chapter 9. We will also need to consider *appropriate* technology for sustainability (and some inappropriate ones), as we do in Chapter 10.

This book is not a book written for academics, it seeks to demystify 'sustainability' for the 'educated layperson' who wants to delve deeper into what 'sustainability' means, or *should mean*. It seeks to cut through the jargon and theory so as to help people consider the real problems that hold us back from reaching sustainability, and consider the real solutions that can break through these barriers. It is for those who want to find out how we can act to bring into being an ecologically sustainable society and future. We shall look at how to meaningfully define 'sustainability' (and its strands). We shall look at what economic sustainability should be. We shall look at the key importance of ecological sustainability, arguably the most over-looked strand of the topic. We shall also look at what social sustainability should be. Like most other scholars, I too conclude that we *need all three* strands of sustainability, but with the terms actually defined, and the key solutions stated and acted on.

The final chapter looks at the solution frameworks we need in order to sustain the Earth and a sustainable future. Along the way, we shall discover that key things are 'broken'. In the West (now globalised widely) we shall see that our economy is broken, our society is broken, and our ecosystems are breaking. It is thus first and foremost the task of 'sustainability' to *heal* these broken things into a harmonious and sustainable whole. That is the 'Great Work' in front of us (Berry 1999) that we discuss in Chapter 11. It is a challenge, but also an opportunity to get things right. We will find in this book that again and again we discover that the main thing holding humanity back from reaching a meaningful sustainability is our blind commitment to endless growth. Growth in numbers of people, growth in the GDP, and growth in consumption and use of resources. The key and inescapable conclusion of three decades of environmental science is that endless growth is the *cause* of unsustainability. However, this is a reality that society remarkably rarely acknowledges or discusses. What we need now is a growth in key intangible things, a growth in accepting the reality of the predicament we face, a growth in compassion and caring, and finally a rapid growth in action towards real solutions, not tokenism.

In this book I shall refer to 'Nature' with a capital 'N'. It is worth explaining that this is not because I 'deify' Nature, make a fetish of it, or believe in any human/Nature 'dualism'. It is simply a mark of respect for the 'more-than-human world' (Abram 1996) that has nurtured humanity physically and spiritually since we evolved. Given the generosity and beauty that Nature has manifested to humanity, it is a respect I believe is fully deserved. Sustainability at its core (I shall argue) is the search to find a harmony between human culture and the Nature from which we evolved and of which we are a part. And reaching that harmony and balance is something well worth striving for: the true path to wisdom.

References

Abram, D. (1996) *The Spell of the Sensuous*, New York: Vintage Books (Random House).
Berry, T. (1999) *The Great Work: Our Way into the Future*, New York: Bell Tower.
Engelman, R. (2013) 'Beyond sustainababble', in *State of the World 2013: Is Sustainability Still Possible?*, ed. L. Starke, Washington: Island Press.
EPA (2013) 'Teaching stewardship and sustainability', US Environmental Protection Agency, online, available at: www.epa.gov/region8/ee/teachingsustainability.html (accessed 26 January 2013).
Foster, J. (2008) *The Sustainability Mirage: Illusion and Reality in the Coming War on Climate Change*, London: Earthscan.
GFN (2013) 'World footprint, do we fit on the planet?', Global Footprint Network, online, available at: www.footprintnetwork.org/en/index.php/GFN/page/world_footprint/ (accessed 31 August 2013).
Harris, S. and Throsby, D. (1998) 'The ESD process: background, implementation and aftermath', in *The ESD Process: Evaluating a Policy Experiment*, eds. C. Hamilton and D. Throsby, Canberra: Australian Academy of Sciences, pp. 1–19.
MacNeill, J. (2006) 'The forgotten imperative of sustainable development', *Environmental Policy and Law*, vol. 36, no. 3/4, pp. 167–170.
MEA (2005) 'Ecosystems and human well-being: opportunities and challenges for business and industry', Millennium Ecosystem Assessment, online, available at: www.millenniumassessment.org/documents/document.353.aspx.pdf (accessed 26 January 2013).

Ott, K. and Doring, R. (2008) 'Strong sustainability and environmental policy: justification and implementation', in *Sustaining Life on Earth: Environmental and Human Health through Global Governance*, ed. C. Soskolne, New York: Lexington Books.

Soskolne, C., Westra, L., Kotze, L, Mackey, B., Rees, W. and Westra, R. (2008) 'Global and local contexts as evidence for concern', in *Sustaining Life on Earth: Environmental and Human Health through Global Governance*, ed. C. Soskolne, New York: Lexington Books.

Sukhdev, P. (2010) 'Preface' to *The Economics of Ecosystems and Biodiversity: Ecological and Economic Foundations*, ed. P. Kumar, London: Earthscan.

WCED (1987) *Our Common Future*, World Commission on Environment and Development, London: Oxford University Press.

1
THE 'OLD' SUSTAINABILITY
A story of listening and harmony

> *It tells me*
> *That actually to care*
> *To listen,*
> *To love the land*
> *To revere harmony*
> *Always was, always will be*
> *In the vast scheme of things*
> *So ethically and spiritually . . .*
> Sane.
>
> *(From 'Madness', Haydn Washington 2013)*

'Sustainability' did not spring forth new-born in 1987 in the report 'Our Common Future' (WCED 1987). What we *mean* by sustainability today draws on a long history of people thinking (and feeling) about living in harmony and balance with Nature, and recording this story as lore or 'law' (Knudtson and Suzuki 1992). However, one could be excused for thinking this is not the case, as almost every book and article on sustainability starts with 'Our Common Future' in 1987 (WCED 1987), or possibly with Rachel Carson's seminal book 'Silent Spring' in 1962. Indeed there is a curious blindness in academia as to the underpinnings that form the foundation for the common parlance of 'sustainability'. Many books will not mention it, while others will touch on it briefly in passing (Edwards 2005). The philosophical, ethical and historical roots that support ideas of harmony and sustainability are not immediately obvious. They are worthy of explanation and acknowledgment and hence form the first chapter of this book. This history, I believe, is essential to understanding the thinking and the thrust of concern within society as to 'why' we want and need to be sustainable, as to why we should bother to *care*.

The new terms 'sustainability' or 'sustainable development' are after all quite prosaic and dry-as-dust. In themselves they are not inspiring, they don't engage people's creativity and imagination. And they don't in fact explain the deep ethical and spiritual concerns that have driven people for thousands of years to speak of how we should 'live in harmony' with Nature. 'Harmony' and 'balance' are essential words here, ones we don't use much regarding sustainability today. They have fallen out of favour with science in academia, in part because they are deemed 'unscientific', and partly (as we shall see in Chapter 5) because 'balance' is criticised by some ecological theorists. The old 'sustainability' is in fact a lot broader than the new sustainability, so broad that many would argue it is not 'sustainability' but something else, either philosophy or environmental ethics. Perhaps it may even be described disparagingly as 'tree-hugging'. However, I believe the old sustainability is actually critical to understanding *why* people argue for and believe in sustainability. It is especially important to help us demystify the term, to help us think about what sustainability 'should be', if we are to be successful and turn our environmental problems around so as to reach a sustainable future. The old sustainability underpins the need to discuss 'worldview' and ethics when one talks about 'sustainability'. How do we see our relationship to Nature, are we part of it or are we its masters? How do we deal ethically with Nature? Does it have 'rights', and do we have a responsibility to respect and care for it. These are considered in detail in Chapter 8.

I come to the understanding of the 'old' sustainability as both an environmental scientist and a long-time conservationist (and bushwalker) working on wilderness issues in Australia. I did my PhD 'The Wilderness Knot' (Washington 2006) on the confusion and tangled meanings around the term 'wilderness' (see Wilderness Truths website, online, available at: www.wildernesstruths.com). So I have read widely the works of those who loved the land and argued for its protection, for its long-term sustainability in the sense that it would still be natural in the future. A number of scholars report that 'conservation' is seen as being separate from environmentalism (e.g. Harding *et al.* 2009), the latter of which is deemed to be focused more on pollution and biodiversity loss. That is why some argue that environmentalism started in 1962 with the publication of 'Silent Spring', which dealt with pesticide pollution. Perhaps conservation and environmentalism do have a different focus (Foreman 2012), but that doesn't mean that conservation has not had a huge influence on our interest about 'sustainability'. Yet this distinction is in many ways artificial, first because large natural areas (aka 'wilderness' as defined by IUCN 2008) are necessary to clean up pollution and maintain ecosystems, ecosystem services and biodiversity (Mackey *et al.* 1998). The supposed separation is a false distinction, since sustainability as a concept is built on the foundation of how people relate to Nature, and that has been influenced strongly by writers and conservationists around the world for hundreds (indeed thousands) of years. To understand sustainability we need to look at the history of human interaction with the land, and the ideas and ideologies that affect this. People's desire to protect Nature (conservation) is thus central to understanding 'sustainability'.

8 The 'old' sustainability

The broadness in the old meaning of sustainability is about how we as a society continue to live in balance and harmony with Nature into the future, so that *both* continue to exist as dynamic, creative entities. I should explain up front that 'balance' as a term is open to co-option, just as 'sustainability' has been. Many developers argue for 'balance' each time a development proposal comes forward. In their PR spin, Nature (which provides essential ecosystem services for our society) has 'too much land and resources' and the supposed 'balance' is to give more to society (read developers) for their particular project. This fundamentally misunderstands how ecosystems work and that ecosystem services and processes underpin our society (see Chapter 5). Such a twisted idea of 'balance' has led to the ongoing cumulative impact or 'death by a thousand cuts' that increasingly whittles down ecosystems to the point of collapse. The 'balance' that past and present indigenous cultures understood was an acknowledgment that humans should not take too much, that they must work with the natural world in harmony (e.g. Neidjie *et al.* 1985). As we shall see, such a balance in the past was based on respect. This was a 'sacred balance' (Suzuki and McConnell 1999).

The idea of humans living in harmony and balance with Nature is not something 'new' that we have just 'discovered', this is something humanity (or at least its thinkers and 'feelers') has always pondered. In fact, this is the 'Wisdom of the Elders' that goes back to pre-history (Knudtson and Suzuki 1992). We have an environmental crisis precisely *because* we (as a society) have forgotten the old sustainability and the teachings and wisdom of millennia. The old sustainability speaks of terms such as: harmony, balance, reverence, sacredness, spirituality, respect, care, witness, responsibility, custodianship, stewardship, beauty and even *love*. These terms rarely appear in the new and sanitised 'sustainable development'. In fact some have labelled this as 'neo-environmentalism', the situation where people have moved to speak only of 'carbon footprints', 'natural resources' or 'energy audits' and cease to speak about *caring for the land* (Kingsnorth 2013). That is a great pity to my mind.

The old sustainability may still grant us an inkling about how humanity, especially modern society, can live in harmony with Nature in the long-term. If we reconnect with the old sustainability again, then arguably this might have value in judging between the many conflicting views regarding 'sustainability' we are deluged with today.

A history of *caring*

The old sustainability considers things such as: harmony, beauty, responsibility, guardianship, custodianship, co-existence, values, humility, ethics and worldviews. In this sense 'sustainability' has been passed down in oral traditions (e.g. Neidjie *et al.* 1985) and is seen in the writings where people recorded their thoughts and feelings. This idea of harmony and respectful co-existence with (and reverence for) Nature has been around probably for as long as humans have existed. We see signs of such reverence even in the waterfall ceremonies of chimpanzees (Goodall 2013). It can be seen in the idea of the Magna Mater, the 'Great Mother', that can be seen in the incredible beauty and loving artistic expression present in pre-historic art (Oelschlaeger 1991). Indeed, some anthropologists describe this ancient human worldview as *Homo spiritualis*, that humanity was centrally focused on its spiritual relationship with Nature (Herzog 2010). Of course today it could be said we are *Homo economicus* (Daly and Cobb 1994, p. 89), but it is worth understanding we were not always so, and to consider whether we need to remain so.

The old sustainability can be seen in the lore and 'law' of traditional cultures, in the 'Wisdom of the Elders' (Knudtson and Suzuki 1992). It can be seen in the surviving stories and poetry of many cultures (Washington 2002). This respect and reverence for Nature is thus a continuing well-spring within our cultures that has underpinned the idea of 'sustainability' that has re-emerged in recent decades. It is thus important we understand the roots of the term, that 'sustainability' gives expression to a feeling that most human cultures have espoused: humanity must live in harmony with Nature. This feeling of being part of Nature, of believing Nature is sacred, of a feeling that we must maintain the harmony of the Universe, has emerged again and again in cultures across the world (O'Hanlon 2012). I examine below a few of these records that relate to harmony with Nature, and thus underlie the idea of 'sustainability'.

A Guatemalan 'Maya' prayer to the guardian of the forest, Pokohil expresses a joyous gratitude towards Nature, and the sense of only taking what you need (Knudtson and Suzuki 1992):

> O Pokohil, today you have shown favour,
> And have given some of your beasts, some of your deer.
> Thank you Pokohil
> See, I bring you flowers for your deer.
> Perhaps you have counted them.
> Two of them are missing;
> They are the ones the Old One (the hunter) caught,
> You gave them to him.

The eleventh-century Chang Tsai placed this inscription on the west wall of his office (so it would always be before him) (Berry 1988):

> Heaven is my father
> and earth is my mother,
> and even such a small creature as I
> finds an intimate place in its midst.
> That which extends throughout the universe,
> I regard as my body
> And that which directs the universe
> I regard as my nature.
> All people are my brothers and sisters
> And all things are my companions.

The Great Law of the North-American Hodenosaunee people (Josephy Jr 1995) says:

> Whenever the statesmen of the League shall assemble they shall offer thanks to the earth where men dwell, to the streams of water, the pools and the lakes, to the maize and the fruits, to the medicinal herbs and trees, to the forest trees for their usefulness, and to the animals that give their pelts for clothing.

Black Elk of the Sioux evoked the unity of life (Knudtson and Suzuki 1992):

> It is the story of all life that is holy and is good to tell, and of us two-leggeds sharing in it with the four-leggeds and the wings of the air and all green things; for these are children of one mother and their father is one spirit.

Luther Standing Bear (1928) of the Dakota spoke of when a young Dakota man went off on a vision quest. During such a quest one spends several days naked and fasting in the mountains. In a prayer to the Universe, the Dakota asked:

> O Wakan-Tanka, grant that this young man may have relatives; that he may be one with the four winds, the four Powers of the world, and with the light of the dawn. May he understand his relationship with all the winged peoples of the air.... Our Grandmother and Mother (Earth) ... this young man wishes to become one with all things.... For the good of all your peoples, help him!

Standing Bear explains that the Dakota made sure children knew that wherever they went, they would be greeted by the warm, reassuring presence of the local life, geological features and natural forces, which were often as trusted, familiar and communicative with them as members of their families back home. Through this process, Dakota children came to an early understanding 'that we are of the soil and soil of us', that 'we love the birds and beasts that grew with us on this soil', and that a bond existed between all living things because they all 'drank the same water and breathed the same air' (Standing Bear 1928).

Aboriginal Elder Bill Neidjie (1985) in 'Kakadu Man' talks movingly of the land:

> This ground and this earth ... like brother and mother. Trees and eagle ... you know eagle? He can listen. Eagle our brother, like dingo our brother. We like this earth to stay, because he was staying for ever and ever. We don't want to lose him. We say 'Sacred, leave him' ... our story is in the land ... it is written in those sacred places. My children will look after those places, that's the law.

David Mowaljarlai, a Ngarinyin elder of the Kimberleys in Western Australia writes of his bonding with the land (Tacey 2000):

> You have a feeling in your heart that you're going to feel your body this day, get more knowledge. You go out now, see animals moving, see trees, a river. You are looking at nature and giving it your full attention, seeing all its beauty. Your vision has opened and you start learning now.... When you touch them, all things talk to you, give you their story.... You understand that your mind has been opened to all those things because you are seeing them; because your presence and their presence meet together and you recognise each other. These things recognise you. They give their wisdom and their understanding to you when you come close to them

The 'Statement of Hopi religious leaders' from the American southwest says in part (Knudtson and Suzuki 1992):

> This land was granted to the Hopi by a power greater than man can explain. Title is invested in the whole make-up of Hopi life. Everything is dependent on it. The land is sacred and if the land is abused, the sacredness of Hopi life will disappear and all other life as well.

Knudtson and Suzuki (1992) in 'Wisdom of the Elders', when speaking of human bonding to Nature, observe: 'Its ancient Laws remain timeless, eternally binding human beings to live in harmony with and respect for other species'. Similarly within Western culture, Nature has long been venerated. The nineteenth-century poet Gerard Manley Hopkins (1918) wrote in 'God's Grandeur' that 'There lives the dearest freshness deep down things', that Nature has wondrous powers of regeneration.

So the older and broader view of sustainability came out of a deep bond and love of the land that human cultures have *always* felt. It came from a sense of responsibility and guardianship to protect that land. Rather than speaking about 'sustainability', people spoke of 'harmony', 'balance' and 'respect'. They spoke of responsibility to the land. It is this long-term (if suppressed) cultural belief in the need to keep the sacred 'balance of Nature' (Suzuki and McConnell 1999) that I believe underlies today's interest in and support for 'sustainability'. It is thus important we understand the deep roots that lie behind the modern term, especially since they are often overlooked.

Modernism and the decline of harmony

The idea of a balance of Nature and living in harmony with the land has taken a battering from the ideology of *modernism*. 'Modernism' is central to how humans treat Nature today. It is a historical movement that 'begins with the Renaissance and extends to the present' (Oelschlaeger 1991, p. 68). Modernism continued the humanisation of wild Nature initiated by the early agriculturists, and operated through science, technology and liberal democracy. It consists of several processes that intertwine, being the Renaissance, the Reformation, the Enlightenment and the democratic, industrial and scientific revolutions (ibid. p. 68). Modernism arguably underlies the emergence of a profound anthropocentrism still dominant in the world, where Nature is conceived of as 'nothing more than matter-in-motion' (ibid. p. 69).

The Renaissance brought forward the idea of the secular 'State', while the Reformation proclaimed the central place of the individual. Humans increasingly looked through 'economic rather than religious spectacles' and the consumer society lay just around the corner (Oelschlaeger 1991, p. 74). No aspects of modernism have had a greater effect than 'science' and 'economics'. Galileo's new science, Bacon's new logic, Descartes' mechanistic reductionism and Newton's physics were central. They represent a shift in worldview so radical 'that the very meaning of the word nature was changed' from an organism to a mechanistic paradigm (ibid., pp. 76–77). 'Nature' in effect became an object of scientific study, and the idea of Nature as animate and living was replaced with the idea of a cold and lifeless machine. Descartes proposed that mind (*res cogitans*) and matter (*res extensa*) are distinct, that the natural world is a machine (Godfrey-Smith 1979, Abram 1992). Newton gave a logical and ostensibly absolute understanding of the natural world, where natural change was reduced to an illusory status, being rather the mechanical repetition of predictable phenomena (Oelschlaeger 1991, pp. 85–89). This approach has also been described as the Linnaean or 'imperial' approach in ecology, though Worster (1994) notes that within ecology there was also another stream of thought, being the 'Arcadian' or naturalist approach. It should not be forgotten that this other naturalist stream does still exist, as arguably it is this view that we see emerge in the scientists who *do* speak out on behalf of Nature (for example, Ehrlich 1986; Suzuki and Dressel 1999). It is also the view of other non-scientist scholars (e.g. Berry 1988; Collins 2010; Macy 2012).

Economics is another key component of modernism (see Chapter 4). Adam Smith (1776) in 'The Wealth of Nations' set in motion 'that modern shrine to the Unattainable: infinite needs' (Oelschlaeger 1991, p. 92). Unlimited growth was the ethical justification for capitalism. Consumption and never ending growth were deemed to be good in modernism, which completed 'the intellectual divorce of humankind from nature'. It draws 'a boundary between an objective or scientific and a poetic or aesthetic view of nature' (Oelschlaeger 1991, pp. 95–98).

The revolt against modernism

The Romantic poets (such as Wordsworth and Coleridge) revolted against modernism. The Romantic writers valued an immediate personal relationship to Nature. Wordsworth (1888) in 'The Tables Turned' wrote words that can be read as a response to 'reductionist' science:

> Sweet is the lore which Nature brings;
> Our meddling intellect
> Mis-shapes the beauteous forms of things:—
> We murder to dissect.
> Enough of Science and of Art;
> Close up those barren leaves;
> Come forth, and bring with you a heart
> That watches and receives.

Romantics saw Nature as alive, created by divine providence, and the idea of 'mere matter' was sterile to them (Oelschlaeger 1991, p. 99). Romanticism can thus be understood as an aesthetic reaction against mechanistic materialism. For romantics, scientific Nature was devoid of taste, sight, sound and feeling, while 'poetic' Nature was alive, subjective, an aesthetic delight. Oelschlaeger (1991, p. 113) notes:

> the Romantics were concerned with affective immediacy: they followed a direct intuitive path to a realisation of the unity of nature ... the Romantic poets are not tender-hearted nature lovers but address issues of fundamental philosophical import – concerns central to the nineteenth-century idea of nature and humankind's relation to it.

Others built on the Romantic poets' re-establishment of the old idea of sustainability. The first great writer was Henry David Thoreau, whose book 'Walden: or Life in the Woods' (1854) is about his time living alone near a lake in Concord, Massachusetts. In 'Sounds', Thoreau (1854) writes:

> I sat in my sunny doorway from sunrise till noon, rapt in a reverie amidst the pines and hickories and sumachs, in undisturbed solitude and stillness, while the birds sang around or flitted noiseless through the house ... I grew in those seasons like corn in the night, and they were far better than any work of the hands would have been. They were not time subtracted from my life, but so much over and above my usual allowance.... Instead of singing like the birds, I silently smiled at my good fortune.

John Muir (in Oelschlaeger 1991) was a nineteenth-century wilderness campaigner who wrote eloquently about Nature:

> When I entered this sublime wilderness the day was nearly done, the trees with rosy, glowing countenances seemed to be hushed and thoughtful ... and one naturally walked softly and awe-stricken among them. I wandered on, meeting nobler trees where all are noble, subdued in the general calm, as if in some vast hall pervaded by the deepest sanctities and solemnities that sway human souls.

Another later eloquent speaker for the land was Aldo Leopold, the father of the 'Land Ethic'. In 'Sand Country Almanac' (Leopold 1949) he wrote:

> The land ethic simply enlarges the boundaries of the community to include soils, waters, plants, and animals, or collectively: the land. This sounds simple: do we not already sing our love for and obligation to the land of the free and the home of the brave? Yes, but just what and whom do we love?... A land ethic of course cannot prevent the alteration, management, and use of these 'resources,' but it does affirm their right to continued existence, and, at least in spots, their continued existence in a natural state.

Later again, a philosophical movement 'deep ecology' developed, led by Norwegian Arne Næss (1984), who argued: 'The well-being of non-human life on Earth has value in itself. This value is independent of any instrumental usefulness for limited human purposes'. Philosopher and theologian Thomas Berry (1988) in his illuminating book 'The Dream of the Earth' concluded:

> The difficulty presently is with the mechanistic fixations in the human psyche, in our emotions and sensitivities as well as in our minds. Our scientific inquiries into the natural world have produced a certain atrophy in our human responses. Even when we recognize our intimacy, our family relations with all the forms of existence about us, we cannot speak to those forms. We have forgotten the language needed for such communication. We find ourselves in an autistic situation. Emotionally, we cannot get out of our confinement, nor can we let the outer world flow into our own beings. We cannot hear the voices or speak in response.

And yet, despite this, Berry noted that: 'We are constantly drawn toward a reverence for the mystery and the magic of the Earth and the larger Universe with a power that is leading us away from our anthropocentrism'.

The Nature writings of Thoreau, Muir, Leopold, Berry and other Nature-writers thus re-stated the 'old sustainability' belief in respect for Nature and its harmony and balance. Their writings collectively ask us to recall that we used to understand that Nature had intrinsic value, that it was something sacred we should respect and care for. We discuss intrinsic value and worldview further in Chapter 8.

Conclusion

Arguably the dominance of modernism within Western society (and spread around the world by globalisation) has served to bury and deny the old idea of sustainability. Rather than something 'sacred' we need to work in harmony with, Nature became just a machine, something to be dissected, used and 'mastered' by the anthropocentric mind as part of a techno-centric future. However, in the end reality has reasserted herself, because natural limits are *real*. A real environmental crisis has been the result. One that modernism caused because it forgot or denied the old sustainability.

Academia (and business and governments too) may overlook the broader, older meaning of sustainability (as they busily jump through various academic and ideological hoops). However, those seeking to demystify sustainability cannot afford to ignore the old meaning of sustainability. If the community at large supports 'sustainability', this is largely because of the old broad meaning based on caring for Nature. It is this meaning that underlies the modern terms 'sustainable development' and 'sustainability'. Those were the roots onto which the modern terms have been grafted, though many seem to overlook this. Next, we need to examine the key events of the 1960s and 1970s (Chapter 2) and then the rise of the 'new' sustainability (Chapter 3).

References

Abram, D. (1992) 'The mechanical and the organic: on the impact of metaphor in science', *Wild Earth*, vol. 2, no. 2, pp. 70–75.
Abram, D. (1996) *The Spell of the Sensuous*, New York: Vintage Books (Random House).
Berry, T. (1988) *The Dream of the Earth*, San Francisco: Sierra Club Books.
Collins, P. (2010) *Judgment Day: the Struggle for Life on Earth*, Sydney: UNSW Press.
Daly, H. and Cobb, J. (1994) *For the Common Good: Redirecting the Economy Toward Community, the Environment, and a Sustainable Future*, Boston: Beacon Press.
Edwards, A. (2005) *The Sustainability Revolution: Portrait of a Paradigm Shift*, Canada: New Society Publishers.
Ehrlich, P. (1986) *The Machinery of Nature*, New York: Simon and Schuster.
Foreman, D. (2012) *Take Back Conservation*, Colorado: Raven's Eye Press.
Godfrey-Smith, W. (1979) 'The value of wilderness', *Environmental Ethics*, vol. 1, pp. 309–319.
Goodall, J. (2013) 'Chimpanzee central – waterfall displays', Jane Goodall Institute, online, available at: www.janegoodall.org/chimp-central-waterfall-displays (accessed 21 August 2013).
Harding, R., Hendriks, C. and Faruqi, M. (2009) *Environmental Decision-making: Exploring Complexity and Context*, Sydney: Federation Press.
Herzog, W. (2010) 'Cave of forgotten dreams', a documentary film by Werner Herzog on Chauvet Cave that contains some of the oldest human paintings, online, available at: www.wernerherzog.com/index.php?id=64.
Hopkins, G. M. (1918) *Poems of Gerard Manley Hopkins*, London: Humphrey Milford.
IUCN (2008) IUCN Guidelines for Applying Protected Area Management Categories. Edited by Nigel Dudley. International Union for Conservation of Nature, Gland, Switzerland.
Josephy Jr, A. M. (1995) *500 Nations: an Illustrated History of North American Indians*, New York: Hutchinson/Pimlico.
Kingsnorth, P. (2013) 'Dark ecology: searching for truth in a post-green world', *Orion Magazine*, January/February 2013, online, available at: www.orionmagazine.org/index.php/articles/article/7277.
Knudtson, P. and Suzuki, D. (1992) *Wisdom of the Elders*, Sydney: Allen and Unwin.
Leopold, A. (1970 [1949]) *A Sand Country Almanac, with Essays on Conservation from Round River*, New York: Random House.

Mackey, B., Lesslie, R., Lindenmayer, D., Nix, H. and Incoll, R. (1998) *The Role of Wilderness in Nature Conservation (Report to Environment Australia)*, Centre for Research and Environmental Studies, Canberra: Australian National University.

Macy, J. (2012) 'A wild love for the world', Joanna Macy interview by Krista Tippett, 'On Being', American Public Media, 1 November, online, available at: www.onbeing.org/program/wild-love-world/61 (accessed 11 September 2013).

Næss, A. (1984) 'A defence of the deep ecology movement', *Environmental Ethics*, vol. 6, p. 266.

Neidjie, B., Davis, S. and Fox, A. (1985) *Kakadu Man: Bill Neidjie*, Queanbeyan, Australia: Mybrood.

O'Hanlon, E. (2012) *Eyes of the Wild: Journeys of Transformation with the Animal Powers*, Washington: Earth Books.

Oelschlaeger, M. (1991) *The Idea of Wilderness: from Prehistory to the Age of Ecology*, New Haven and London: Yale University Press.

Smith, A. (1776) 'An inquiry into the nature and causes of the wealth of nations', in *Great Books of the Western world*, ed. R. Hutchins (1952), vol. 39:1, Chicago, IL: Encyclopedia Britannica.

Standing Bear, L. (1928) *My People the Sioux*, New York: Houghton Mifflin.

Suzuki, D. and Dressel, H. (1999) *Naked Ape to Superspecies*, Sydney: Allen and Unwin.

Suzuki, D. and McConnell, A. (1999) *The Sacred Balance: Rediscovering Our Place in Nature*, Sydney: Allen and Unwin.

Tacey, D. (2000) *Re-enchantment: the New Australian Spirituality*, Australia: Harper-Collins.

Thoreau, H. D. (1995 [1854]) *Walden: or Life in the Woods*, New York: Dover Publications.

Washington, H. (2002) *A Sense of Wonder*, Sydney: Ecosolution Consulting.

Washington, H. (2006) 'The wilderness knot'. PhD Thesis, Sydney: University of Western Sydney, online, available at: http://arrow.uws.edu.au:8080/vital/access/manager/Repository/uws:44.

Washington, H. (2013) *Poems from the Centre of the World*, online, available at: www.lulu.com/au/en/shop/haydn-washington/poems-from-the-centre-of-the-world/paperback/product-21255751.html.

WCED (1987) *Our Common Future*, World Commission on Environment and Development, London: Oxford University Press.

Wordsworth, W. (1888) 'The tables turned: an evening scene on the same subject' in *Wordsworth: the Complete Poetical Works*, London: Macmillan, stanzas 7 and 8 quoted in text.

Worster, D. (1994) *Nature's Economy: a History of Ecological Ideas*, Cambridge: Cambridge University Press.

2
THE 1960S TO THE PRESENT
Key conferences and statements

> *We must join together to bring forth a sustainable global society founded on respect for nature, universal human rights, economic justice, and a culture of peace. Towards this end, it is imperative that we, the peoples of Earth, declare our responsibility to one another, to the greater community of life, and to future generations.*
>
> *(Preamble, Earth Charter 2000)*

The background to the emergence of 'sustainability' and 'sustainable development', terms now commonplace, is rooted in the period from the 1960s through to 'Our Common Future' (WCED 1987). It is important to understand this history behind the emergence of the terms 'sustainable development' and 'sustainability'. A key background is the history of key conferences and statements on the environment.

The fabulous 1960s and 1970s

It is not possible to pin down a precise 'starting point' for the emergence of modern environmentalism, or what Rolston (2012, p. 20) has called 'the environmental turn'. The previous chapter looked at the historical and cultural underpinnings of the 'old' sustainability', which gave birth to the conservation movement. There is however a debate about whether 'environmentalism' derived from conservation, or whether they are separate (Foreman 2012). Some argue it was pollution that started environmentalism (e.g. pesticides), and some sustainability scholars do not mention the legacy of visionary scholars such as Thoreau and Muir, and even Leopold and his 'Land Ethic'. As Chapter 1 discussed, this omission is a shame, for the conservation legacy in Western society has contributed directly to the meaning of the term 'sustainability' that emerged into common use in 1987.

Many scholars agree that several milestones in sustainability took place in the United States in the 1960s and 1970s, which led to the 'environmental revolution' (e.g. Barrow 1999). The 1960s in the United States was a period that saw the questioning of the 'status quo' (e.g. the Vietnam War, the status of women and African Americans in the United States, the US civil rights movement). At the same time, a number of factors led some of the public to question humanity's treatment of the environment. This was fuelled by events such as the spillage of 70,000 tons of crude oil from the tanker 'Torrey Canyon' in 1967 (Barkham 2010), together with other pollution events. While there had been people such as Thoreau (1854), Marsh (1864), Muir (1916) and Leopold (1949) who had written about Nature, more and more people in the 60s and 70s then started to write about environmental issues, including pollution, population growth and ecosystem degradation.

Prominent among these were the writings of Rachel Carson, a biologist with the US Fisheries and Wildlife division. Her most famous book 'Silent Spring' was published in 1962, and provided an alarming account of the effects of persistent organochlorine pesticides (such as DDT) on wildlife, especially birds. Written for the layperson, it was widely read, and communicated the importance of 'ecology' as a discipline for understanding human impacts on the natural world. Al Gore (1992, p. 3) the former US vice-president, recalls: 'My mother was one of many who read Carson's warnings and shared them with others. She emphasised to my sister and me that this book was different – and important. Those conversations made an impression'. Carson was vilified by the pesticide industry as both hysterical and wrong. Recently, deniers of the environmental crisis have even taken to calling Carson a 'mass murderer', ostensibly because she stopped the use of DDT, which controlled mosquitoes and hence malaria (Tren and Bate 2000; Delingpole 2009). This seems to have been with the aim of discrediting Carson, for 'Silent Spring' galvanised the public recognition that we *need to regulate* environmental problems, and hence regulate industry. Deniers seemed to hope that if Carson was discredited, then perhaps the regulation of business would be abandoned (see Chapter 9). In fact, DDT was abandoned mainly because mosquitoes *evolved resistance* against it (Oreskes and Conway 2010). History and careful biological research have shown that Carson was dead right (Van Emden and Peakall 1999; Swartz 2007).

A flood of literature on problems in the environment followed in the late 1960s and through the 1970s, which listed the damage of modern society to the environment. Much of this literature came from biologists and ecologists who sought to inform the public about the impact modern society was having on Nature. Some warned that unless humans curbed their exponentially growing impact, then the world's ecosystems would degrade and perhaps collapse, impacting on human society. The early emphasis was on 'pollution', particularly its effects on wildlife. 'Ecology' was suddenly a term that was commonly in use. From Carson and other authors such as Paul Ehrlich (1968) and Barry Commoner (1971) came the strong message that humans are a 'part of Nature' and not 'apart from it'. This came to be part of a new (or perhaps rediscovered) worldview that has been called *ecocentrism* (see Chapter 8).

The Californian ecologists, professors Paul and Anne Ehrlich, have been among the most influential writers on environmental science since the late 1960s. They address the relations between population, resources, economic activity, technology and the environment. Paul Ehrlich's (1968) most famous book is 'The Population Bomb'. The Ehrlichs and John Holdren coined the equation $I = PAT$, or environmental impact equals population × affluence × technology (Ehrlich *et al.* 1977). Our impact on the Earth is thus the number of people times their affluence (per capita consumption of resources) times the technology we use. A lively debate in the literature emerged in the early 1970s between the so-called 'neo-Malthusians' and the 'Cornucopians'. 'Neo-Malthusianism' generally refers to people with similar concerns as Thomas Malthus, who argued for population control programmes to ensure resources were retained for current and future populations (Marsh and Alagona 2008). The 'neo-Malthusians' in the 1960s and 1970s were mainly environmental scientists writing about environmental problems, who predicted that if we didn't act then the Earth's systems would collapse from overuse of resources and associated pollution. They were labelled 'prophets of doom' by their opponents (e.g. Maddox 1972). These opponents described themselves as 'optimists', while others called them 'Cornucopians' or techno-centrists. A *Cornucopian* is a futurist who believes that continued progress and material wealth will be met by continued advances in technology. Fundamentally, they believe that there is enough matter and energy on the Earth to provide for an ever-rising population and consumption. The term comes from the *Cornucopia*, the 'horn of plenty' of Greek mythology, which magically supplied its owners with endless food and drink. The Cornucopians do not see resource depletion as limiting human progress, and see technology as 'saving the day' through better technology and substitution of resources. A related term is 'techno-centrism', which sees technology as solving all problems. A key Cornucopian champion in the 1970s was John Maddox, then editor of the journal 'Nature' (Maddox 1972). More recently, economist Julian Simon (1998) has championed Cornucopianism.

In 1972, the book 'Limits to Growth' pointed out that our population growth and increasing consumption of resources would exceed planetary limits around the middle of the twenty-first century, most likely leading to societal collapse, though action could avert this (Meadows *et al.* 1972). It was commissioned by the 'Club of Rome', a think tank founded in 1968 of 'citizens sharing a common concern for the future of humanity'. The book used the 'World3' computer model to simulate the consequence of interactions between the Earth's systems and humanity. Five variables were examined: world population, industrialisation, pollution, food production and resource depletion. Two of the scenarios modelled showed 'overshoot and collapse' of the global system by the mid to latter part of the twenty-first century, while a third scenario resulted in a 'stabilised world'. This report was strongly criticised by Cornucopians and neoclassical economists seeking to deny the environmental crisis, who predictably labelled the authors 'prophets of doom'. However, a 30-year review of the 'standard' model in the 'Limits to Growth' has shown that it has been *remarkably accurate* (Turner 2008). Wijkman and Rockstrom

(2012, p. 161) note eight other studies that conclude the same. The latest version of the 'Limits to Growth' came out in 2004 (Meadows *et al.* 2004) and confirms earlier forecasts, but does show a scenario for reaching sustainability, providing we abandon denial and solve key problems (such as our addiction to growth).

Around this time, the space programme provided the first pictures of the Earth hanging in space. The fact that the Earth was a 'closed system' materially, with finite resources, was stressed. The parallel of the Earth, with the astronauts in their 'spaceship' having to provide for all their needs and dispose of all their wastes within this closed system, did not go unnoticed. The concept of 'Spaceship Earth' (Boulding 1966) further reinforced the popular environmental writings of the time.

Key conferences and reports

The 1972 Stockholm conference: In 1972 the UN Conference on the Human Environment was held in Stockholm, Sweden. This was the first world environment conference ever held. Each member country reported on its environmental situation and the measures to control problems. The main product of the conference was the Stockholm Declaration of 26 principles, intended as a foundation for future development. There was also an 'Action Plan for the Human Environment' of 109 recommendations, which covered species conservation, forests, pollution, development and trade. It set up a new organisation: UNEP (The United Nations Environment Programme). Brenton (1994, p. 50) summarised the Stockholm Conference:

> It is difficult to view Stockholm as much more than a cosmetic event. It demonstrated to Western publics that their governments were taking the international environment seriously. It exposed developed and developing countries to the gulf between their respective views of the environmental issues and demonstrated the impossibility of conducting global discussions on environmental problems without also facing the developmental issues linked to them.... On the other hand, Stockholm was the first full-scale display of the new environmental diplomacy, conducted largely in the open with intense media and NGO attention. The mere fact of the conference also forced many governments to focus seriously on environmental issues in a way that they had not done before.

The 'North' was mainly concerned about pollution and ecosystem decline, but the 'South' was more concerned about poverty, disease and hunger. The South thus tended to see economic development as the solution. As we shall discuss in Chapter 4, arguably in the North this was the *cause* of the problem. This divergence of views has continued to the present day.

The World Conservation Strategy (1980): The World Conservation Strategy (WCS 1980) was launched in 1980 by the IUCN (International Union for Conservation of Nature and Natural Resources), UNEP and WWF (the World Wildlife Fund). It introduced not only the concept of 'sustainable development' but also the term 'sustainable' in relation to human use of the biosphere. It stated: 'We have not inherited the earth from our parents, we have borrowed it from our children'. This was a powerful expression of environmental concern at the end of the 1970s. The above has links with the idea of 'inter-generational equity' which was a key element of the definition of sustainable development, which we discuss next under 'Our Common Future'.

The aim of the World Conservation Strategy was: 'To help advance the achievement of sustainable development through the conservation of living resources'. And the three main objectives of the strategy were listed as:

a   Maintain essential ecological processes and life support systems
b   Preserve genetic diversity
c   Ensure the sustainable utilisation of species and ecosystems.

The World Conservation Strategy argued that rather than 'conservation' and 'development' being mutually exclusive activities (as generally argued by conservationists up to that time), these were 'interdependent'. The WCS argued that development *requires* the conservation of the living resource base on which it ultimately depends, as in the longer term development will not be able to take place unless we conserve our living resources. Likewise, conservation will not occur unless at least minimal standards of development are met, i.e. basic needs of food, shelter and clean water. The debate as to whether these are really 'interdependent' (and whether this claim has justified continuing business-as-usual) still continues (discussed further later).

'Our Common Future' – The Brundtland Report (WCED 1987): This report played a key role in achieving world-wide attention for the concept of 'sustainable development'. The World Commission on Environment and Development, chaired by Gro Harlem Brundtland (former Norwegian prime minister), produced a report 'Our Common Future' (WCED 1987). Its most quoted definition of 'sustainable development' was: 'Development that meets the needs of the present without compromising the ability of future generations to meet their own needs'.

The report highlighted three fundamental components to sustainable development: environmental protection, economic growth and social equity. Meadows *et al.* (2004, p. 123) note that 'Our Common Future' labelled world trends simply 'unsustainable' but did not find it politically opportune to say 'the human world is beyond its limits', much less grapple seriously with the question of what to do. The concept of 'sustainable development' focused attention on finding strategies to promote economic and social advancement in ways that avoid environmental degradation, over-exploitation or pollution. It has been said that it 'sidelined less productive debates about whether to prioritise development or the environment' (SD2015 — 2014).

Brenton (1994, p. 129) noted:

> The genius of the piece lies in its adoption and promulgation of the concept of 'sustainable development' ... [which] effectively bridged the intellectual and political gap which had been apparent at least since Stockholm between those (particularly in the developing world) arguing for economic growth, and those (particularly in the developed world) arguing for environmental protection. It encouraged growth, but incorporated environmental concern in order to ensure that growth should not ultimately undo itself. In one neat formula, Mrs Brundtland had provided a slogan behind which first world politicians with green electorates to appease, and third world politicians with economic deprivation to tackle, could unite. The formula was of course vague, but the details could be left for later.

And indeed, the details *have* been left for later – effectively we are still waiting. This report contained a statement of eight principles known as the Tokyo Declaration, the first of which was to 'revive economic growth' (see Chapter 4). 'Our Common Future' recommended economic growth of 5 per cent in order to alleviate poverty, though it did argue for a change in the 'quality of growth' so that 'sustainability, equity, social justice and security are firmly embedded'. Duchin and Lange (1994, pp. 5–8) carried out an 'input–output' analysis of the report, concluding 'the position taken in the Brundtland Report is not realistic' as their analysis shows that 'the economic and environmental objectives of the Brundtland Report cannot be achieved simultaneously'. In other words, given environmental limits, endless growth is not sustainable. This is a key reality we will return to repeatedly as we try to demystify 'sustainability'.

The Earth Summit, 1992: The United Nations Conference on Environment and Development met in Rio de Janeiro in June 1992. It was commonly called the 'Earth Summit', and was the culmination of the discussions on sustainability placed on the international agenda by 'Our Common Future'. Although it has been branded a failure by many, others have stressed the significance of the meeting and the gains made. The Earth Summit was the largest gathering of world heads of government, and the most complex model of environmental negotiation that has taken place: 178 countries participated, represented by 120 heads of government. There were 12,000 official delegates, 2,000 bureaucrats, 1,400 accredited non-government organisations and 7,000 accredited journalists. In addition to the 'official' conference, over 20,000 delegates attended the 'Global Forum', a gathering of non-government organisation representatives. The major outcomes from the conference were the Rio Declaration, Agenda 21 and the Convention on Biodiversity and the Framework Convention on Climate Change (later called the Kyoto Protocol).

The Rio Declaration was a general statement of 27 principles providing guidance for the conduct of nations (and their inhabitants) in relation to environment and development. It was not a legally binding document. It was quite anthropocentric, for principle 1 stated: 'Human beings are at the centre of concerns for sustainable development'. It was also pro-growth, as principle 12 stated: 'States should cooperate to promote a supportive and open international economic system that would lead to economic growth and sustainable development in all countries'. However, the Rio Declaration also defined the Precautionary Principle (principle 15) as: 'Where there are threats of serious or irreversible damage, lack of full scientific certainty shall not be used as a reason for postponing cost-effective measures to prevent environmental degradation'.

This has become a key part of sustainability and environmental protection within legislation. There has been criticism of this principle and the Rio+20 Earth Summit reviewed the Precautionary Principle as a result of the wide range of interpretations and debates it has engendered. However, it still remains of paramount importance to the future of sustainability in business and elsewhere as a means of inducing developers and innovators to reflect deeply on what they are doing (Kopnina and Blewitt 2015). The Rio Declaration was said to be the 'rudder' for the 'ship of action' that was 'Agenda 21' (UN 1992). This is a vast document of 40 chapters and 115 programmes, which some saw as the 'Action Plan' for the twenty-first century. It aimed for the introduction of sustainability into all nations. It argued for acceptance of a 'responsibility' by higher income countries for less developed countries and their needs. It also recognised the role of public participation by all governments. These points and the Precautionary Principle were important steps towards sustainability.

However, Agenda 21 has been criticised as lacking sufficient strength in relation to issues such as population, consumption patterns in the North, and the role of transnational corporations (Foster 2003; SD21 2012). The question of the lasting impact of Agenda 21 remains open. It seems to have faded somewhat in the last few years in terms of public commitment by governments. Certainly, it has sadly not become the 'ship of action' that moves society to a sustainable future. The negotiations leading up to the summit have been said to have had a profound influence in putting 'sustainability' on the political agenda in nations where this had been lacking. It required a vast environmental education process, as government, business and the community had to grapple with this concept. We shall discuss later the question of whether such a learning process has actually led to permanent change, or whether it has resulted in a tokenism that has allowed the continuation of a business-as-usual approach.

At the Earth Summit a new body was formed, the Sustainable Development Commission, with the task of reviewing the work of the UN 'family' regarding implementation of Agenda 21. At the summit, the business community saw the UN continuing to play a strong role regarding the environment, and this influenced national policies (Willums and Golüke 1992). Expansion of the Global Environment Facility (set up in 1990 to assist developing countries with environmental problems) was endorsed in order to fund activities leading to sustainability. It recommended that nations provide 0.7 per cent of GDP to achieve this. However, a shortfall in funding is common (GEF 2013).

Sachs (2002) summarised concerns about the 'failure' of the Earth Summit:

> Rio 1992 reveals itself a vain promise. While governments at the Earth Summit had committed themselves in front of the eyes and ears of the world to curb environmental decline and social impoverishment, no reversal of these trends can be seen a decade down the line. On the contrary the world is sinking deeper into poverty and ecological decline, notwithstanding the increase of wealth in specific places.... Fifty years from now, when the Earth is likely to be hotter in temperature, poorer in diversity of living beings, and less hospitable to many people, Rio might be seen as the last exit missed on the road to decline.

The Earth Charter, 2000: We should also discuss the Earth Charter (2000), though this is not a legally binding document, of all these formal documents it is the most widely endorsed as a charter of sustainability. It is worth reading because of its scholarship and breadth of scope. 'Our Common Future' called for a 'new charter' to set new norms to guide the transition to sustainable development. Following that, discussion about an 'Earth Charter' took place leading to the Earth Summit in 1992, but timing was not right to complete this then. The 'Rio Declaration' became the statement of the achievable consensus at that time. In 1994, Maurice Strong, secretary general of the Rio Summit, and Mikhail Gorbachev, former president of the USSR, launched an initiative to develop an 'Earth Charter' as a civil society initiative. The initial drafting and consultation process drew on hundreds of international documents. An independent Earth Charter Commission was formed in 1997 to oversee the development of the text, analyse the outcomes of a world-wide consultation process, and to come to agreement on a global consensus document.

The Earth Charter Commission came to a consensus on the Earth Charter in March 2000 at a meeting in Paris. The Earth Charter is now increasingly recognised as a global consensus statement on the 'meaning of sustainability', the challenge and vision of sustainable development, and the principles by which sustainable development is to be achieved (see Earth Charter in Action website, online, available at: www.earthcharterinaction.org). Soskolne (2008) argues it is the key sustainability document, which should be used as a 'covenant' to protect the natural world we depend on. The charter was also an important influence on the Plan of Implementation for the UNESCO Decade for Education on Sustainable Development. The preamble of the Earth Charter states: 'We stand at a critical moment in Earth's history, a time when humanity must choose its future. As the world becomes increasingly interdependent and fragile, the future at once holds great peril and great promise'.

Unlike earlier documents, which ignored the intrinsic value of Nature, Principle 1a is to: 'Recognise that all beings are interdependent and every form of life has value regardless of its worth to human beings'. It is worth emphasising here that many scholars (author included) believe that the Earth Charter is probably as close as we have come to 'agreement' on the vision and principles that form the backbone of a meaningful sustainability (Soskolne 2008). It is a truly visionary document that combines compassion for humanity *and* Nature, and argues for justice for both. It is probably the best international document we have to help demystify sustainability.

Johannesburg Summit, 2002: The World Summit on Sustainable Development took place in Johannesburg, South Africa in September, 2002. It was convened to discuss sustainable development by the United Nations. The Johannesburg Declaration (see UN Documents website, online, available at: www.un-documents.net/jburgdec.htm) was the main outcome. It has been widely argued that this summit failed to live up to the expectations of the 1992 Earth Summit, and failed to address climate change and reform global environmental governance (e.g. La Vina *et al.* 2003; Foster 2003).

Rio+20 Summit, 2012: There are divergent views about the success of this, the most recent summit. It certainly made little impact compared to the Earth Summit 20 years previously. Some saw the outcomes as positive (Ivanova 2014). Some 592 world leaders and representatives were present, represented by 57 heads of state and 31 prime ministers. However, heads of state from the United Kingdom, Germany and the United States were absent. Two decades after the initial Earth Summit, the environmental problems addressed were still seen as valid (i.e. had not been solved). Two main topics dominated, first the transition to a 'green economy' as a key to achieving sustainable development and poverty eradication, and second working towards strengthening the institutional framework for sustainable development at all levels (TBB 2012). Regarding the 'green economy' the 'Common Vision' states: 'We also reaffirm the need to achieve sustainable development by: promoting sustained, inclusive and equitable economic growth'. We discuss this further in Chapter 4.

Due to the then 'global financial crisis', the conference focused on the economic aspect of environmental protection, leaving less time for climate change (TBB 2012). The only real outcome document was 'The Future We Want' (see UN website, online, available at: www.un.org/en/sustainablefuture/), which was not binding. Greenpeace described this as 'the longest suicide note in history', while Ivanova (2014, p. 142) notes that in the document the world 'environment' has almost disappeared. Delegates also agreed to accept new 'Sustainability Development Goals' to replace existing Millennium Development Goals in 2015. Barry (2014, p. 150) notes that it remains to be seen if the new goals will represent business-as-usual or provide something new by recognising the biophysical limits of economic growth. The conference did not result in any significant agreements or set targets for pressing issues such as climate change or overpopulation. For these reasons, Rio+20 has been said to be an 'epic failure' (Montague 2012) or disappointment (Wijkman and Rockstrom 2012, p. 185) that produced 'lots of gaseous talk and no significant action' (Prugh 2013, p. 111). Barbara Stocking, the chief executive of Oxfam, noted:

> Rio will go down as the hoax summit. They came, they talked, but they failed to act. Paralysed by inertia and in hock to vested interests, too many are unable to join up the dots and solve the connected crisis of environment, equality and economy.
>
> *(Montague 2012)*

Daniel Mittler, political director of Greenpeace concluded:

> The epic failure of Rio+20 was a reminder that short-term corporate profit rules over the interests of people.... They spend $1 trillion a year on subsidies for fossil fuels and then tell us they don't have any money to give to sustainable development.
>
> *(Montague 2012)*

In summary, the above events, conferences and statements set the scene for the rise of the 'new' sustainability we discuss next.

References

Barkham, P. (2010) 'Oil spills: legacy of the Torrey Canyon', *Guardian*, 24 June 2010, online, available at: www.guardian.co.uk/environment/2010/jun/24/torrey-canyon-oil-spill-deepwater-bp (accessed 25 February 2013).

Barrow, C. (1999) *Environmental Management for Sustainable Development* (second edition), New York: Routledge.

Barry, E. (2014) 'A policy mechanism for ensuring sustainable development', in *State of the World 2014: Governing for Sustainability*, ed. L. Mastny, Washington: Island Press.

Boulding, K. (1966) 'The economics of the coming spaceship Earth', in *Environmental Quality in a Growing Economy*, ed. H. Jarrett, pp. 3–14. Baltimore, MD: Resources for the Future, Johns Hopkins University Press.

Brenton, T. (1994) *The Greening of Machiavelli: The Evolution of International Environmental Politics*, London: The Royal Institute of International Affairs and Earthscan.

Carson, R. (1962) *Silent Spring*, Boston: Houghton Mifflin.

Commoner, B. (1971) *The Closing Circle: Nature, Man, and Technology*, New York: Knopf.

Delingpole, J. (2009) 'Rachel Carson, environmentalism's answer to Pol Pot', *Telegraph*, online, available at: http://blogs.telegraph.co.uk/news/jamesdelingpole/9917408/Rachel_Carson_environmentalisms_answer_to_Pol_Pot/ (accessed 24 February 2013).

Duchin, F. and Lange, G. (1994) *The Future of the Environment: Ecological Economics and Technological Change*, New York: Oxford University Press.

Earth Charter (2000) The Earth Charter, Earth Charter Initiative, online, available at: www.earthcharterinaction.org/content/pages/Read-the-Charter.html (accessed 11 April 2013).

Ehrlich, P. (1968) *The Population Bomb*, New York: Ballentine Books.

Ehrlich, P., Ehrlich, A. and Holdren, J. (1977) *Ecoscience: Population, Resources, Environment*, New York: W. H. Freeman and Co.

Foreman, D. (2012) *Take Back Conservation*, Colorado: Raven's Eye Press.

Foster, J. (2003) 'A planetary defeat: the failure of global environmental reform', *Monthly Review*, vol. 54, no. 8, online, available at: http://monthlyreview.org/2003/01/01/a-planetary-defeat-the-failure-of-global-environmental-reform (accessed 4 March 2013).

GEF (2013) Summary of 43rd Meeting of the Global Environment Facility Council, online, available at: www.iisd.ca/gef/council43/ (accessed 20 August 2013).

Gore, A. (1992) *Earth in the Balance: Ecology and the Human Spirit*, New York: Plume.

Ivanova, M. (2014) 'Assessing the outcomes of Rio+20', in *State of the World 2014: Governing for Sustainability*, ed. L. Mastny, Washington: Island Press.

Kopnina, H. and Blewitt, J. (2015) *Sustainable Business: Key Issues*, London: Routledge.

La Vina, A., Hoff, G. and DeRose, A. (2003) 'The outcomes of Johannesburg: assessing the World Summit on Sustainable Development', *SAIS Review*, vol. 23, no. 1, pp. 53–70.

Leopold, A. (1970 [1949]) *A Sand Country Almanac, with Essays on Conservation from Round River*, New York: Random House.

Maddox, J. (1972) *The Doomsday Syndrome*, New York: McGraw-Hill.

Marsh, G. P. (1864) *Man and Nature; or, Physical Geography as Modified by Human Action*, New York: Scribner.

Marsh, M. and Alagona, P. (2008) *Barrons AP Human Geography*, 2008 edition, New York: Barron's Educational Series.

Meadows, D., Meadows, D., Randers, J. and Behrens, W. (1972) *The Limits to Growth*, Washington: Universe Books.

Meadows, D., Randers, J. and Meadows, D. (2004) *Limits to Growth: the 30-year Update*, Vermont: Chelsea Green.

Montague, B. (2012) 'Analysis: Rio+20 – epic failure', Bureau Stories, online, available at: www.thebureauinvestigates.com/2012/06/22/analysis-rio-20-epic-fail/ (accessed 29 February 2013).

Muir, J. (1916) *A Thousand Mile Walk to the Gulf*, Boston: Houghton Mifflin.

Oreskes, N. and Conway, M. (2010) *Merchants of Doubt: How a Handful of Scientists Obscured the Truth on Issues from Tobacco Smoke to Global Warming*, New York: Bloomsbury Press.

Prugh, T. (2013) 'Getting to true sustainability', in *State of the World 2013: Is Sustainability Still Possible?*, ed. L. Starke, Washington: Island Press.

Rolston III, H. (2012) *A New Environmental Ethics: the Next Millennium for Life on Earth*, New York: Routledge.

Sachs, W. (2002) 'Ecology, justice and the end of development', in *Environmental Justice, Discourses in International Political Economy – Energy and Environmental Policy*, ed. J. Byrne, vol. 8, New Brunswick and London: Transaction Publishers, pp. 19–36.

SD2015 (2014) The Brundtland Report 'Our Common Future', online, available at: www.sustainabledevelopment2015.org/AdvocacyToolkit/index.php/earth-summit-history/historical-documents/92-our-common-future (accessed 21 July 2014).

SD21 (2012) *Review of Implementation of Agenda 21 and the Rio Principles: Synthesis*, Stakeholder Forum for a Sustainable Future, online, available at: http://sustainabledevelopment.un.org/content/documents/641Synthesis_report_Web.pdf (accessed 25 February 2013).

Simon, J. (1998) *The Ultimate Resource 2*, New Jersey: Princeton University Press.

Soskolne, C. (2008) 'Preface' to *Sustaining Life on Earth: Environmental and Human Health through Global Governance*, ed. C. Soskolne, New York: Lexington Books.

Swartz, A. (2007) 'Rachel Carson, mass murderer?: The creation of an anti-environmental myth'. *Fair: Fairness and Accuracy in Reporting*, September 2007, online, available at: http://fair.org/extra-online-articles/rachel-carson-mass-murderer/ (accessed 24 February 2013).

TBB (2012) 'RIO+20 Earth Summit summary', The Trend is Blue, online, available at: www.thetrendisblue.com/article.cms/rio20-earth-summit-summary.

Thoreau, H. D. (1995 [1854]) *Walden: or Life in the Woods*, New York: Dover Publications.

Tren, R. and Bate, R. (2000) 'When politics kills: malaria and the DDT story', Competitive Enterprise Institute, online, available at: http://cei.org/PDFs/malaria.pdf (accessed 24 February 2013).

Turner, G. (2008) 'A comparison of the *Limits to Growth* with 30 years of reality', *Global Environmental Change*, vol. 18, no. 3, pp. 397–411.

UN (1992) *Agenda 21*, United Nations Conference on Environment and Development,

June 1992, online, available at: http://sustainabledevelopment.un.org/content/documents/Agenda21.pdf (accessed 15 June 2014).

Van Emden, H. and Peakall, D. (1999) *Beyond Silent Spring: Integrated Pest Management and Chemical Safety*, London: UNEP/ICIPE, Chapman & Hall.

WCED (1987) *Our Common Future*, World Commission on Environment and Development, London: Oxford University Press.

WCS (1980) *World Conservation Strategy: Living Resource Conservation for Sustainable Development*, IUCN/UNEP/WWF, online, available at: www.a21italy.it/medias/31C2D26FD81B0D40.pdf (accessed 21 July 2014).

Wijkman, A. and Rockstrom, J. (2012) *Bankrupting Nature: Denying our Planetary Boundaries*, London: Routledge.

Willums, J. and Golüke, U. (1992) *From Ideas to Action: Business and Sustainable Development*, The ICC report on the greening of enterprise 92, Report commissioned by the International Environmental Bureau of the International Chamber of Commerce, Oslo: ICC Publishing.

3
RISE OF THE 'NEW' SUSTAINABILITY – THE WEAK AND THE STRONG

> *I also never thought that the concept of SD could and would be interpreted in so many different ways ... many of them of course are totally self-serving. I no longer shock easily but to this day I remain stunned at what some governments in their legislation and some industries in their policies claim to be 'sustainable development'. Only in a Humpty Dumpty world of Orwellian doublespeak could the concept be read in the way some suggest...*
>
> (Jim MacNeill 2006, pp. 3–4)

We have seen that the events of the 1960s and 1970s and 1980s led to the common use of the words 'sustainability' and 'sustainable development'. It is time to dig deeper and examine the rise of the 'new' sustainability and address the contested question of whether sustainability and sustainable development are in fact the *same thing*, and whether sustainability should be 'weak' or 'strong'.

Destination or journey?

The first part of addressing the key question of whether 'sustainability' equals 'sustainable development' is to discuss the 'destination or journey' debate. This lies at the core of why sustainability has been said to be a 'contested concept' (Jacobs 1999; Harding *et al.* 2009). Many other important concepts are also contestable, such as democracy, justice or freedom, though this doesn't mean we stop using such terms. Is 'sustainability' a journey or a destination? Submissions to the Australian Sustainability Charter (CoA 2007) were divided on this, as are academics to this day. Many argue that sustainability is the 'ultimate goal' or destination, while 'sustainable development' is the framework and path through which sustainability is achieved. Others argue that 'sustainable development' is an oxymoron, as we discuss later.

30 Rise of the 'new' sustainability

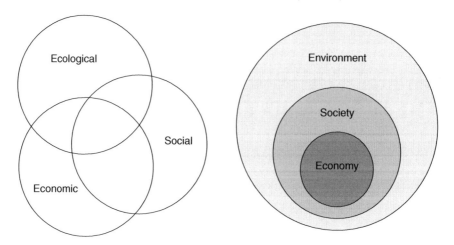

FIGURE 3.1 The 'sustainability trinity'. On the left is the traditional three-ring model, while on the right is a more ecologically realistic model of how they relate, as the economy is part of society, which is part of the natural world.

Clearly, there will always need to be ongoing improvement to any sustainability strategy over time. Hence if we have a goal of 'sustainability' there will have to be a *transition process* to get there. In this transition there must be flexibility and adaptability to handle new issues and improve sustainability in the light of new developments. The question here is whether there is *real value* in assigning the term 'sustainability' to the 'destination' and the term 'sustainable development' to the 'journey'? Are there not many terms in our society that are both destination and journey? 'Conservation' is a goal *and* a process to get there, and nobody considers this a problem. Democracy and justice similarly. If we are to make a distinction of sustainability being the 'destination' and the transition process to get there being called 'sustainable development', then we should look closely at the meaning of the term sustainable *development*.

Is 'sustainable development' a code for sustainable *growth*?

This is a critical question, one surprisingly rarely asked (or answered). Indeed the denial around this question is very relevant to Chapter 9. Giddens (2009) has noted that semantically 'sustainability' implies continuity, stability and balance, while 'development' implies dynamism and change. Gowdy (2014, p. 39) notes that many well-meaning efforts in sustainability have gone off track 'by trying to reconcile sustainability with the dominant ideology of growth and accumulation'. So ... do people mean by 'sustainable development' that there will be continued growth in GDP, population and resource use, albeit described sometimes as 'sustainable growth' or 'green growth'? Elliott (2013, pp. 25–39) summarises the changing perceptions of 'development' over time. In the 1950s and 1960s development was unequivocally *economic growth* via modernisation and Westernisation. Rostow

(1960) proposed 'stages of development' to transform a traditional society via increased mining and manufacturing to mass consumption. 'Dependency theory' (Frank 1967) depicted the industrialised nations as keeping the developing nations underdeveloped. Elliott argues that the 1980s saw more arguments for 'growth with equity'. The 1990s were dominated by a neoliberal 'magic of the market' approach (Simon 2008). The point here is that all these approaches equated 'development' with economic growth.

Elliott (2013, p. 38) argues that the 2000s (and the 'global financial crisis') brought forward a campaign for anti-globalisation and a questioning of 'neoliberalism'. Neoliberalism is a modern politico-economic theory favouring free trade, privatisation, minimal government intervention in business and reduced public expenditure on social services (Kopnina and Blewitt 2015). However, while a questioning of neoliberalism may be occurring in some circles, neoliberal market approaches still remain dominant, as noted in the 2014 State of the World Report (Mastny 2014). The history and discourse on 'development' and development studies has thus shown a continuous support for *economic growth* being the key to development (Wijkman and Rockstrom 2012).

This confirms the views of several authors (e.g. Bartlett 1996) who argue that 'sustainable development' is often seen as being based on continuing economic growth. It is illuminating in this regard to consider the foreword to 'Our Common Future' itself (WCED 1987). Cavagnaro and Curiel (2012, p. 78) make it clear that 'Our Common Future' took a *techno-centric* stance, which would 'make way for a new era of economic growth' (WCED 1987, p. 8). This was in part due to the mandate the UN gave to WCED. 'Our Common Future' states on p. xii: 'What is needed now is a new era of economic growth – growth that is forceful and at the same time socially and environmentally sustainable'. It does, however, acknowledge later on p. 9: 'Thus sustainable development can only be pursued if population size and growth are in harmony with the changing productive potential of the ecosystem'.

Principle 1 of the Tokyo Declaration that accompanied 'Our Common Future' was to 'Revive growth'. In fact it is a little recognised fact that 'Our Common Future' argued that 'sustainable development' required a global GDP annual growth rate of 5 per cent (where the GDP would double in just 14 years). This was argued as being necessary to solve poverty (rather than tackle the thorny problem of redistribution). However, Bartlett (1996) points out that there was at the same time no implicit recognition in the document that this would impact on ecological limits. As noted in Chapter 2, Duchin and Lange (1994, pp. 5–8) have noted that realistically 'the economic and environmental objectives of the Brundtland Report cannot be achieved simultaneously'. In other words, you cannot protect the environment (and ecosystem services) if you plan to endlessly grow human numbers and resource use.

At the Earth Summit in 1992, the Rio Declaration Principle 12 stated that 'States should cooperate to promote a supportive and open international economic system that would lead to *economic growth and sustainable development* in all countries' (my emphasis). Similarly, 20 years later, the 'Common Vision' of the Rio+20 summit in 2012 similarly states: 'We also reaffirm the need to achieve sustainable development by: promoting sustained, inclusive and *equitable economic growth*' (my emphasis). As Chapter 4 discusses, UNEP's (2011) 'Green Economy' was also seen as a 'new engine of growth'. Similarly, the OECD (2011) published a document 'Towards Green Growth' to 'devise new ways of ensuring that the growth and progress we have come to take for granted are assured in the years to come'. All these statements by major international organisations on 'sustainable development' have helped to embed the idea that sustainable development and economic growth *must go together*.

This trend continues today in modern textbooks on sustainability. Braungart and McDonough (2008, p. 11) in 'Cradle to Cradle' argue their book 'goes beyond the environmental chorus saying that growth is wrong'. In fact they argue for 'good growth', though this seems mostly to be in intangibles such as health and education. Edwards (2010) gives great examples of positive actions one can take, but when speaking of 'thriveability' doesn't really confront or deal with the issue of endless growth, and supports a green (but growth) economy. Khalili (2011) states: 'Sustainability, at its core, means sustaining a sustainable economy … sustainable development can be warranted, and social sustainability can be linked to environmental sustainability and sustained economic growth'.

Cavagnaro and Curiel (2012) state in the first paragraph of their introduction that: 'The ultimate goal of sustainable development is securing a better quality of life for all, both now and for future generations, by pursuing responsible economic growth'. Similarly, Elliott (2013, p. 185) sees optimism in sustainable development from new institutions for the 'revitalisation of economic growth' albeit with 'greater weight given to environmental concerns'.

A number of writers (e.g. Bonnett 1999) argue that a large part of the appeal of 'sustainable development' has been that it seems to bring into harmony the idea of sustaining what is valued (Nature), while still accommodating human aspirations to develop (grow). This raises the suspicion that it involves a 'certain sleight of hand' (Bonnett 1999). Shiva (1992) suggests that to the Western mind 'development' cannot escape market connotations as being seen as economic growth. Combining 'development' with 'sustainability' thus holds the danger of changing the fundamental motive to conserve (conservation) into one of 'finding substitutes', which feeds the underlying motive for material growth and consumption. Shiva concludes that from the perspective of a market economy 'sustainable development' will inevitably be measured (and come to mean) the maximisation of profits and capital accumulation. Rist (1997) argues that trading on the ambiguities of 'sustainable development' has allowed policy-makers to give the impression that they want to sustain Nature, while in fact focusing on economic growth. Bonnett (1999) concludes that the great danger in the sustainable development debate is that it 'becomes a term of political convenience used to mask and/or legitimate vested interests'. Rees (2010) points to the 'myopic futility of a global "development" model based on perpetual growth on a finite planet'. Wijkman and Rockstrom (2012, p. 156) conclude that 'green growth' is a myth that can give false hope and be an excuse, where nothing fundamental changes. They also conclude that the growth model is undermining our happiness and causing a 'social recession'.

Kopnina (2012) points out that education for 'sustainable development' does not fully explain the impact of economic development on the ecological health of the biosphere. She concludes:

> The real danger of education for sustainable development is that it confuses the teacher and the student about inherent contradictions of the 'having your cake and eating it' approach. The most fundamental paradox of SD can be summed up in its oxymoronic goal of both promoting development through economic growth and re-distribution of wealth and keeping the health of the ecosystem – including humans – intact.

Kopnina and Blewitt (2015) conclude that education for sustainable development reflects the dominant neoliberal discourse with its emphasis on continuing to promote open markets, the equitable distribution of wealth and social justice. They note also that aside from The Earth Charter initiative, few programmes aimed at education for sustainable development explicitly address the paradoxes of sustainable development or consider the long-term implications of expanding population and wealth. So, clearly there has been a long history of 'sustainable development' going hand in hand (and indeed being seen to *require*) economic growth (see Chapter 4). This is a central consideration for the question asked below.

Does 'sustainable development' *equal* sustainability?

Terborgh (1999) notes that 'sustainable development' is seldom rigorously defined, and thus without definition means anything anyone wants it to mean. He concludes that the proliferation of formulaic solutions has led to confusion and disillusionment with the whole concept. Meadows *et al.* (2004, p. 254) believe it is simplest to say that a sustainable society is one that can 'persist over generations', one that is far-seeing enough, flexible enough and wise enough not to undermine either its physical or it social systems of support. Of course if that was true, it could not be based on endless growth. Rolston (2012, p. 37) concludes that sustainable development is 'more charming than meaningful', as it can be twisted to fit any ongoing worldview the term risks being co-opted by those who wish to perpetuate the expansionist (growth) model, only now pretending they are doing what can be done forever. Following 'Our Common Future' (WCED 1987), the two terms ('sustainable development' and 'sustainability') have generally been spoken of interchangeably (e.g. Harding *et al.* 2009; Cavagnaro and Curiel 2012). Accordingly, many people still tend to equate sustainability with sustainable development, without really considering if the two are (or should be) the same thing.

'Our Common Future' (WCED 1987) made famous the term 'sustainable development', based on the premise that environmental problems would be solved by further development in a sustainable manner. It argued for environment protection, but also argued for continued (indeed expanded) economic growth. However, the report assumed the two were always compatible, and not at some point *mutually exclusive*. Many environmental scientists and scholars argue the ideology of 'growthism' or 'evermoreism' (Boyden 2004) is in fact the 'root cause' of the environmental crisis and not the way to reach sustainability (Daly 1991; Rees 2008; Jackson 2009). It has been argued that 'sustainable development' is actually an 'oxymoron' or a contradiction in terms (Kopnina 2012), that in many (perhaps most) cases more development *cannot* be sustainable. It has also been argued that it doesn't really assist in Nature conservation (Soulé 2001). With the term 'sustainable development' the focus is on 'development', while with the term sustainability, the focus is on being sustainable – whether that involves development or not.

The key question hinges around what 'development' means. We have seen that for WCED it meant ongoing economic *growth*. Others disagree, as they define 'development' differently. For example, Daly (1991) argues for sustainable development, but only where 'development' is *qualitative* and not quantitative. He points out that on a finite world 'sustainable growth' in population or resource use is indeed an oxymoron. Daly (1991) argues 'sustainable growth' implies an eventual impossibility, which the term 'sustainable development' does not. As Daly (1991) explains: 'If something is non-physical then perhaps it can grow forever, if something can grow forever then certainly it is non-physical'. So if one means by 'development' that society is 'qualitatively' changing (for example developing a better culture) then 'sustainable development' is possible into the future. Hamilton (1997) and Diesendorf (2000) follow this logic. 'Development' in this sense could thus mean intellectual, artistic or cultural development that did not involve logging, mining or urban development. It is the meaning of 'development' in a qualitative sense that can have the attribute of sustainability, not growth. Orr (2003) agrees, noting of 'Our Common Future' that 'Their definition confused sustainable growth, an oxymoron, and sustainable development, a possibility'. Nevertheless, we come back to the fact that the WCED (1987) 'Our Common Future', which coined the almost universal definition of 'sustainable development', *does* indeed rest on an assumption of ongoing growth.

Let us return to *meaning*. I understand why Daly and others have defined 'development' as qualitative, but this is not its common meaning in our society, which clearly means resource use and environmental impact. A 'developer' in common parlance clears land and builds things, he doesn't work on his or her poetry or music. Nor is development taken to be 'qualitative' in the meaning of development in 'Our Common Future', which was powered by a 5 per cent annual growth in GDP. If one uses the normal meaning of 'development', which involves resource use and environmental impact, then I am forced to conclude that 'ecologically' sustainable development is indeed almost always an oxymoron, as in most instances any further development that causes environmental impact is now *not* ecologically sustainable. This is in fact precisely why we have an environmental crisis, because 'sustainable growth' is fundamentally an oxymoron, a very damaging one. On a finite world endless physical growth cannot be sustainable. Sustainability now arguably means no more growth, and possibly 'degrowth' (at least for the developed world), as Chapter 4 discusses.

Victor (2008) observed that the absence of a completely unambiguous definition of sustainable development in WCED (1987) helped make it possible for governments and businesses to adopt a goal of sustainable development without compromising their adherence to economic growth (see Chapter 4). Today there are definitions of 'sustainable development' to suit everyone (Victor 2008). MacNeill (2006, p. 3) was a lead author of 'Our Common Future' and noted:

> I also never thought that the concept of SD could and would be interpreted in so many different ways ... many of them of course are totally self-serving. I no longer shock easily but to this day I remain stunned at what some governments in their legislation and some industries in their policies claim to be 'sustainable development'. Only in a Humpty Dumpty world of Orwellian doublespeak could the concept be read in the way some suggest.... In 1987 we thought the concept was plain enough. We defined it in several ways – ethical, social, ecological ... only one definition grabbed the headlines, however and stuck, unfortunately to the exclusion of all others.

Victor (2008) notes that the Canadian Department of Finance defines sustainable development as 'long term sustainable economic growth based on environmentally sound policies and practices'. This definition ignores any real concept of ecological limits. Talking about 'environmentally sound policies' may be seen as just tokenism to make growth *sound* sustainable.

There is thus a deep confusion about 'sustainable development' and 'sustainability' in terms of their fundamental meaning. 'Sustainability' to my mind must be a much *broader concept* than 'sustainable development'. It should focus on sustainability in the long-term for all aspects of the human and natural environment. It should certainly not be about growth in numbers, resource use or GDP. In our current predicament, with a global ecological footprint of 1.5 Earths (GFN 2014) and a Living Planet index that has dropped by 28 per cent (WWF 2012), ecological sustainability will generally *not* mean development, and certainly cannot mean endless growth. 'Sustainable development' thus seems often to be based on the ideological view of continual 'progress' and endless growth. As Shiva (1992) predicted, 'sustainable development' has been co-opted by market ideology to continue 'business-as-usual' growth. However, this is the *cause* of escalating degradation of the biosphere (see Chapter 5). If we are to demystify 'sustainability' then our focus must (first and foremost) centre on being sustainable, not on endless growth on a finite planet, which is an unsustainable delusion.

We face a critical issue here in terms of what we *mean* by sustainability. Bonnett (2007) argues 'sustainable development' as a term allows for such vagueness that it has enabled the policy-makers and business to give the impression that they are concerned to do one thing – such as sustain natural ecosystems – while in fact attempting something quite different – such as sustain conditions for the continuance of narrowly defined economic growth. We cannot afford to dither on this point or descend into vague generalities (yet society has done this for almost three decades). Otherwise 'sustainability' is in danger of becoming an empty shibboleth (Daly 1991, p. 249). If we are talking about 'sustainability' and a sustainable future, do we believe that the human population can keep growing forever? Do we believe that our use of resources can keep growing forever (see Chapter 7)? Do we believe that we can keep clearing native vegetation endlessly, and destroying ecosystems for human needs? If we don't believe this, then sustainability *cannot* be about development in the sense of growth in a physical sense. If we do believe this, then we are living a delusion, in a fantasy world where we deny ecological reality, simply because that is what we want to believe (see Chapter 9). The scientific data on the environmental crisis is clear, a testament that growth over the last century has *not* been ecologically sustainable (e.g. MEA 2005). Further growth in numbers and resource use will only accelerate the crisis. Despite the hopes of the 'World Conservation Strategy' (WCS 1980) and 'Our Common Future' (WCED 1987), we cannot 'grow' our way out of this crisis. *Growth is the cause, not the cure* (Chapter 4 expands on this). In fact, if you mean growth by the 'development' in sustainable development – as most (including the UN) seem to – then 'sustainable development' in that meaning is actually stopping us reaching a sustainable future. That meaning is not just an oxymoron, it is one of the worst delusions in the list of our denial of the environmental crisis.

This debate remains hotly contested, and some environmental scientists and scholars (whom I respect) disagree on this point. However, this question is the crucial challenge of 'sustainability' today in terms of demystifying it. Sustainability should not be allowed to be subverted and hijacked to justify further 'business-as-usual' growth. This has been the case, and continues. If we are to demystify sustainability, we have to be on the same page and speak of the same meaning. In a finite world, we need to accept once and for all that sustainability *cannot* be about further growth. This challenge remains critical, though still denied. Indeed, I am sure some in academia will disagree with this conclusion.

Sustainability: a 'second wave' of modern environmentalism?

Beder (1996) argues that sustainable development is 'part of a second wave of modern environmentalism'. The 'first wave' was said to be associated with the counter-culture movement of the 1960s and 1970s. Beder states that 'first wave environmentalism' was easily characterised as 'anti-development', and hence was not shared by business. Therefore it was difficult for governments to administer the environmental legislation and policies put in place during the 'first wave'. By contrast she suggests that the 'second wave' of environmentalism 'has had much broader support and has involved governments, business people and economists in the promotion of sustainable development'. She further notes:

> Sustainable development has succeeded in arenas of influence denied to first-wave environmentalism because of its central idea that environmental protection is not necessarily opposed to development. The concept of sustainable development accommodates economic growth, business interests and the free market and therefore does not threaten the power structure of modern industrial societies.
>
> *(Beder 1996, p. xii)*

However, physical growth on a finite planet cannot continue forever. Thus the idea that sustainable development is 'okay' because it allows continuing growth could be seen as not just being tokenism, but co-option. A co-option by those who wish to continue 'business-as-usual' rather than face up to the real problems caused by endless growth, and thus solve the environmental crisis. If by 'development' we mean an increase in population and resource use, then the so-called first wave of sustainability was *absolutely correct* to be anti-development. Beder (2000) also wrote the book 'Global Spin' on the corporate assault on environmentalism, so she well understands that the second wave may often be an exercise in PR, rather than actually a good thing. To acknowledge that the 'concept of sustainable development accommodates economic growth' is to point out that this may not solve environmental problems. If the free market and the consumer economy don't feel threatened by 'sustainable development', then one can validly ask whether this was 'success' and a good thing, or whether this is because the term has been subverted and is part of a denial of the environmental crisis and its root causes? We shall discuss such denial in Chapter 9, and in Chapter 4 discuss the growth assumptions underlying neoclassical economics.

We will also need to consider whether the vaunted 'second wave' has actually been an advance towards a sustainable future, or a retreat from the hard decisions society needs to make. We shall return to this in Chapter 4. A 'third wave' of sustainability is now also spoken of (Benn *et al.* 2014), where 'third wave' corporations are said to reinterpret the nature of a corporation as 'an integrated self-renewing element of the whole society, and in its ecological context' (Dunphy *et al.* 2003). In third wave organisations or 'sustaining corporations', sustainability is said to be 'value-driven' and transformative, responsive to emerging shift in global values. It is also said to make corporate citizenship and corporate sustainability core business strategies. They will respond to the scientific consensus on the environmental crisis, understand the need for international regulation, question the dominant neoliberal economic model (and consider ecological economics), increase their use of the Precautionary Principle, and adopt a 'care for all' approach that considers all humans and Nature also (Benn *et al.* 2014, p. 186). As to whether many corporations have *actually moved* to the third wave, Sloan *et al.* (2013) note: 'Although individual business success stories start to appear, very little information is available on the challenges and intermittent failures firms may encounter on their road towards sustainability'. Harich (2010) argues that the corporation can be seen as a 'new form of life' that actually *blocks* moves to sustainability, and that this will not change until the corporation itself is changed. Sukhdev (2013) lists ways to reform corporations to form a new corporate model (see Chapter 7).

Sustainability – the 'weak', the 'strong' and the 'strongest'

Within academia there has also been much discussion about 'weak' and 'strong' sustainability. To understand these terms, first we need to discuss the meaning of 'natural capital' (also called 'natural assets'). Natural capital is a term used by economists to summarise the multiple services of Nature from which humanity benefits, from natural resources (including food and water), to pollution absorption and environmental amenities (recreation, wildlife tourism, etc.) (Neumayer 2007). Natural capital is thus the extension of the economic notion of 'capital' (manufactured means of production) to goods and services relating to the natural environment. Hence it is the stock of natural ecosystems that yields a flow of valuable ecosystem goods and services for humanity into the future. 'Natural capital' was an idea coined by Vogt (1948) and further developed by Schumacher (1973) in his classic book 'Small is Beautiful', and later by Costanza and Daly (1992). The term is closely identified with economist Herman Daly, ecologist Robert Costanza, and the 'Natural Capitalism' economic model of Hawken *et al.* (1999, but see also Natural Capitalism website, online, available at: www.natcap.org). Hawken *et al.* (1999, p. 2) note:

> Natural capital includes all the familiar resources used by humankind: water, minerals, oil, trees, fish, soil, air, et cetera. But it also encompasses living systems, which include grasslands, savannas, wetlands, estuaries, oceans, coral reefs, riparian corridors, tundras, and rainforests. These are deteriorating worldwide at an unprecedented rate. Within these ecological communities are the fungi, ponds, mammals, humus, amphibians, bacteria, trees, flagellates, insects, songbirds, ferns, starfish, and flowers that make life possible and worth living on this planet.

They go on to observe:

> It is not the supplies of oil or copper that are beginning to limit our development but life itself. Today, our continuing progress is restricted not by the number of fishing boats but by the decreasing numbers of fish; not by the power of pumps but by the depletion of aquifers; not by the number of chainsaws but by the disappearance of primary forests.
> *(Hawken et al. 1999, p. 3)*

Natural capital is an economic metaphor for the limited stocks of physical and biological resource found on Earth (Kumar 2010). This includes non-renewable resources, renewable resources and ecosystem services (see Chapter 5). Ecosystem services demonstrate the significance of ecosystems in providing the 'biophysical foundation' for the development of all human economies. The concept of natural capital implies that the 'savings rate' of an economy is an imperfect measure of what the country is actually *saving*, because it measures only investment in 'man-made' capital. Natural capital is undervalued and sometimes invisible in economic analysis and the GDP. Wijkman and Rockstrom (2012, pp. 129, 170) point out that there is simply not enough natural capital to sustain the policies of conventional growth, and without better management of it, there is no chance of meeting the basic needs of today's population, much less an increasing population.

Finally, we should realise that natural capital is a fairly anthropocentric concept, and 'critical' natural capital even more so. 'Critical natural capital' has been defined as natural capital that is responsible for important environmental functions (for humans) and which cannot be substituted in the provision of these functions by manufactured capital (Ekins *et al.* 2003). In other words they are ecological assets essential for human well-being or survival (Pearce 1993). By using terms such as 'natural resources' and 'natural capital' however, we are effectively reducing the wondrous diversity of life (in all its beauty) down to just a 'resource' and a natural form of 'capital' for neoclassical economics to consider (Rolston 2012). Braungart and McDonough (2008, p. 155) point out that the idea of natural capital might have been valid 200 years ago but 'now it cries out for rethinking'. The point to consider here is that for much of the world's ecosystem services to survive in free market culture, it may *have* to be effectively considered under the neo-classical economic system (see Chapter 4). From an ethical and ecological justice (eco-justice) viewpoint however (Baxter 2005), it is worth remembering that Nature is much *more* than just 'natural capital' (see Chapter 8).

So how can sustainability be *strong* or *weak*? Do we sustain the future or not? The answer to this lies in economic theory, as opposed to ecological reality. Environmental degradation or damage to natural capital has traditionally been seen as an 'externality' by economists. Under this view, you could substitute wealth for destroyed natural capital. In other words if you made enough money, damage to (or destruction of) ecosystems did not matter. The idea of 'weak sustainability' came from the work of economists Solow (1974, 1993) and Hartwick (1978). Solow championed the weak sustainability position that all forms of capital are *interchangeable*. Thus many economists still believe 'weak sustainability' is good enough, where the aggregate stock of manufactured *and* natural assets is not decreasing. This allows substitution of human-made capital for depleted natural capital (Simon 1998).

In the 'weak sustainability view, it is deemed acceptable to pass on a degraded biosphere to the future – as long as we also hand over a lot of money and human assets to them. From an ecological viewpoint, weak sustainability is actually fundamentally *unsustainable*. It is strictly a short-term strategy, as it ignores the ecological underpinnings of how the world really works (Washington 2013). However, it remains the most common approach to sustainability by neoclassical economists and most governments (Hamilton 1997). The commonly used 'cost–benefit analysis' is an instrument of weak sustainability, it requires reduction of all values to monetary values and allows substitution of capital (Rennings and Wiggering 1997). Liberal democracy also typically promotes weak sustainability (Ward 2008).

'Strong sustainability' by contrast recognises the unaccounted (by economists) ecological services and life support functions of natural capital. It requires that natural capital stocks be 'held constant' independently of human-made capital. Daly and Cobb (1994) championed this approach. For proponents of strong sustainability, the substitutability of manufactured for natural capital is seriously limited by such characteristics of natural capital as irreversibility, uncertainty and the existence of 'critical components of natural capital which make a unique contribution to welfare' (Ekins et al. 2003). Thus, no matter how much money you make, it is not acceptable in strong sustainability to degrade natural capital. Strong sustainability is rooted more deeply in ethics (and ecological reality) and has a distinctive focus on natural capital (Soskolne 2008, p. 63). It considers future generations. Strong sustainability has to justify the 'Constant Natural Capital Rule', with only a limited range of substitution of money for natural capital (Ott and Doring 2008, p. 114). From an ecological point of view, strong sustainability makes far more sense if we are to solve the environmental crisis. Strong sustainability does not necessarily mean that *nothing* in Nature can be changed by humans, though Beckerman (1994) portrays it this way (then describing this as 'morally repugnant' if it leads to human misery).

Some authors go further in terms of grades of 'weak' and 'strong'. Khalili (2011, p. 11) describes 'intermediate' sustainability, which requires that 'attention should be given to the composition of that capital from among natural, manufactured and human-made. Efforts should be made to define critical levels of each type of capital, beyond which there is concern about sustainability'. He also describes 'absurdly strong sustainability', where humanity would 'never deplete anything'. He argues 'absurdly, non-renewable resources could not be used at all; for renewable resources only net annual growth rates could be harvested in the form of over-mature portions of the stock'. Solow (1993, p. 180) similarly dismissed the concept that came to be called 'absurdly strong sustainability', saying 'you can't be obligated to do something that is not feasible'. This conveniently ignores the fact that for most of humanity's history, hunter-gatherer societies came very close to this practice. Holland (1997, p. 128) points out that 'absurdly strong sustainability' requires 'some defence of the natural world beyond the call of human interests'. He then asks: 'Why is this supposed to be absurd?'. Some scholars (e.g. Kuhlman and Farrington 2010) argue there is 'a place for both', where strong sustainability indicates what thresholds must not be crossed, while weak sustainability is the yardstick by which policy outcomes within that space are judged. Cavagnaro and Curiel (2012) argue for four types of sustainability, ranging from very weak to moderately weak, moderately strong and very strong.

So where does this leave us? The question I wish to pursue is whether 'strong' sustainability is actually enough, whether it is truly sustainable *in the long term*. Clearly, weak sustainability ignores the fact that humans are dependent on Nature for survival. I will not reiterate the facts behind this as they are stated in Chapter 5, and by the MEA (2005) and I also cover them in my book 'Human Dependence on Nature' (Washington 2013). The idea that natural capital and ecosystem services can be degraded, merely in exchange for 'money', ignores ecological reality. We rely on Nature to survive. It is popular in sustainability circles to speak of humanity and Nature being 'interdependent'. Indeed this goes back to the World Conservation Strategy (WCS 1980). However, this is actually a delusion, Nature doesn't depend on us, *we depend on it*. Weak sustainability is thus an ecological 'dead end', it is not sustainable in terms of sustaining a long-term ecologically viable future for humanity. The economic mantra of substituting money for ecosystem services has been a major contributor to the environmental crisis, and demonstrates the ecological ignorance that permeates our society and mainstream neoclassical economics.

Strong sustainability accepts that natural capital must not be degraded. This makes sense as it accepts ecological limits. However, given we discuss worldview and ethics in Chapter 8, it should be noted that while a vast improvement, strong sustainability does not come from a strong position of 'environmental ethics'. While strong sustainability may appear too 'radical' for neoclassical economists, it is still anthropocentric and narrowly functional. It looks at the minimum biophysical requirements for *human survival*, without regard to other species (Wackernagel and Rees 1996, p. 38). It considers natural capital 'just for the human species'. We do not experience the taste, smell and feel – the sheer sensual exuberance – of Nature just as 'natural capital'. From an ecocentric perspective, one can validly ask whether other species have a right to natural capital also, or is everything just for us?

The preservation of biophysical assets essential to humanity implies the protection of whole ecosystems and thousands of 'keystone species' (see Chapter 5). Given the 'radical uncertainty' (Kumar 2010, p. 38) around *where* the tipping points are where ecosystems collapse, we need to apply the 'Precautionary Principle' to how we interact with Nature. Strong sustainability allows a narrow utilitarian and anthropocentric view of protecting only those ecosystem services we are *aware we need*. However, this may not actually be pragmatic, or even in our human interests in the long-term. It is also certainly not environmentally ethical (Rolston 2012). The most promising hope for maintaining significant biodiversity (and the experience of Nature) under our prevailing value system has been said to be 'ecologically enlightened self-interest' (Wackernagel and Rees 1996). However, even mutual self-interest holds no ground if there are risks, and when exclusive self-interest promises a 'bigger pay off' (Rees 2008, p. 89). As Caldwell (1990) has noted 'Individual self-interest alone will never save the world'. If instead we were to shift to *ecocentric values*, Nature's survival would be ensured even more effectively, along with the ecosystem services we depend on. 'Respect' for (and preservation of) species and ecosystems for their intrinsic and spiritual values would automatically ensure human ecological security as well. The ethical basis of strong sustainability is better than weak sustainability, as we then consider future generations and intergenerational equity. However, it still does not recognise the 'intrinsic value' of Nature.

So perhaps we actually need a *'strongest sustainability'* that accepts the intrinsic value of Nature and is ecocentric? This has rarely been discussed in sustainability literature, but was included briefly in a table by Pearce (1993, p. 19), calling it 'very strong sustainability' and in Cavagnaro and Curiel (2012). Pearce stated it was based on a 'deep ecology' worldview (see Chapter 8) that accepts bioethics and the intrinsic value of Nature. Such a 'strongest' sustainability would in fact most clearly reflect the older, broader meaning of 'sustainability' in terms of harmony and maintaining the balance of Nature (see Chapter 1). Such a 'strongest sustainability' would best describe the 'Great Work' of Earth repair that philosopher Thomas Berry (1999) has called for. We will not reach an ecologically sustainable future in the long-term, I believe, *unless* we actually move to a 'strongest' sustainability. Despite what Solow (1993) claims, a strongest sustainability is anything but absurd. In the long run it is a necessity. Given that we are struggling now just to move away from the ecological unreality of 'weak sustainability' to just get society to accept 'strong' sustainability, it shows we have a lot of work still to do to advance ideas of 'sustainability' to situate it within ecological reality.

Conclusion

The 'new' sustainability is still evolving. The debate about the equivalence of sustainability and sustainable development continues, and this is something that needs to be demystified. I conclude they are *different*, given that development has for so long been equated to growth (which if endless cannot be sustainable). We need now to focus on *sustainability*, not a 'sustainable development' underpinned by endless growth. The weak and strong sustainability debate also continues. However, if we are to solve the environmental crisis, then the ecological flaw in weak sustainability means it cannot be sustainable, even in the medium term. At a minimum we need strong sustainability, and in the long term a 'strongest' sustainability. These debates illustrate the need to finally agree on what we mean when we speak of 'sustainability', and work out how to get there.

A key part of demystification is working out what we mean by the three strands of sustainability – ecological sustainability, economic sustainability and social sustainability. The next three chapters attempt these difficult tasks.

References

Bartlett, A. (1996) 'Reflections on sustainability, population growth and the environment', *Population and Environment*, vol. 16, pp. 5–35.
Baxter, B. (2005) *A Theory of Ecological Justice*, New York: Routledge.
Beckerman, W. (1994) 'Sustainable development: is it a useful concept?', *Environmental Values*, vol. 3, no. 3, pp. 191–209.
Beder, S. (1996) *The Nature of Sustainable Development*, Newham, Australia: Scribe.

Beder, S. (2000) *Global Spin: the Corporate Assault on Environmentalism*, revised edition, Melbourne: Scribe.
Benn, S., Dunphy, D. and Griffiths, A. (2014) *Organisational Change for Corporate Sustainability*, London: Routledge.
Berry, T. (1999) *The Great Work: Our Way into the Future*, New York: Bell Tower.
Bonnett, M. (1999) 'Education for sustainable development: a coherent philosophy for environmental education', *Cambridge Journal of Education*, vol. 29, no. 3, pp. 313–324.
Bonnett, M. (2007) 'Environmental education and the issue of nature', *Journal of Curriculum Studies* vol. 39, no, 6, pp. 707–721.
Boyden, S. (2004) *The Biology of Civilisation: Understanding Human Culture as a Force in Nature*, Sydney: UNSW Press.
Braungart, M. and McDonough, W. (2008) *Cradle to Cradle: Remaking the Way we Make Things*, London: Vintage Books.
Caldwell, L. (1990) *Between Two Worlds: Science, the Environmental Movement and Policy Choice*, Cambridge: Cambridge University Press.
Cavagnaro, E. and Curiel, G. (2012) *The Three Levels of Sustainability*, Sheffield: Greenleaf Publishing.
CoA (2007) 'Sustainability for survival: creating a climate for change, inquiry into a sustainability charter', House of Representatives Standing Committee on Environment and Heritage, Canberra.
Costanza, R. and Daly, H. (1992) 'Natural capital and sustainable development', *Conservation Biology*, vol. 6, pp. 37–46.
Daly, H. (1991) *Steady State Economics*, Washington: Island Press.
Daly, H. and Cobb, J. (1994) *For the Common Good: Redirecting the Economy toward Community, the Environment, and a Sustainable Future*, Boston: Beacon Press.
Diesendorf, M. (2000) 'Sustainability and sustainable development', in, *Sustainability: the Corporate Challenge of the 21st Century*, ed. D. Dunphy, J. Benveniste, A. Griffiths and P. Sutton, pp. 19–37, Sydney: Allen and Unwin.
Duchin, F. and Lange, G. (1994) *The Future of the Environment: Ecological Economics and Technological Change*, New York: Oxford University Press.
Dunphy, D., Griffiths, A. and Benn, S. (2003) *Organisational Change for Corporate Sustainability* (second edition), London: Routledge.
Edwards, A. (2010) *Thriving Beyond Sustainability: Pathways to a Resilient Society*, Canada: New Society Publishers.
Ekins, P., Simon, S., Deutsch, L., Folke, C. and De Groot, R. (2003) 'A framework for the practical application of the concepts of critical natural capital and strong sustainability', *Ecological Economics*, vol. 44, pp. 165–185.
Elliott, J. (2013) *An Introduction to Sustainable Development*, London: Routledge.
Frank, A. (1967) *Capitalism and Underdevelopment in Latin America*, New York: Monthly Review Press.
GFN (2014) 'World footprint: do we fit on the planet?', Global Footprint Network, online, available at: www.footprintnetwork.org/en/index.php/GFN/page/world_footprint/ (accessed 21 July 2014).
Giddens, A. (2009) *Global Politics and Climate Change*, Cambridge: Polity Press.
Gowdy, J. (2014) 'Governance, sustainability, and evolution', in *State of the World 2014: Governing for Sustainability*, ed. L. Mastny, Washington: Island Press.
Hamilton, C. (1997) 'Foundations of ecological economics', in *Human Ecology, Human Economy: Ideas for an Ecologically Sustainable Future*, ed. M. Diesendorf and C. Hamilton, Sydney: Allen and Unwin, pp. 35–59.

Harding, R., Hendriks, C. and Faruqi, R. (2009) *Environmental Decision-Making: Exploring Complexity and Context*, Sydney: The Federation Press.

Harich, J. (2010) 'Change resistance as the crux of the environmental sustainability problem', *System Dynamics Review*, vol. 26, no. 1, pp. 35–72.

Hartwick, J. M. (1978) 'Substitution among exhaustible resources and intergenerational equity', *Review of Economic Studies*, vol. 45 no. 2, pp. 347–354.

Hawken, P., Lovins, A. and Lovins, H. (1999) *Natural Capitalism: Creating the Next Industrial Revolution*, New York: Little Brown and Co.

Holland, A. (1997) 'Substitutability: or why strong sustainability is weak and absurdly strong sustainability is not absurd', in *Valuing Nature; Economics, Ethics and Environment*, ed. J. Foster, London: Routledge.

Jackson, T. (2009) *Prosperity Without Growth: Economics for a Finite Planet*, London: Earthscan.

Jacobs, M. (1999) 'Sustainable development as a contested concept' in *Fairness and Futurity: Essays on Environmental Sustainability and Social Justice*, ed. A. Dobson, Oxford: Oxford University Press.

Khalili, N. (ed.) (2011) *Practical Sustainability: From Grounded Theory to Emerging Strategies*, London: Palgrave Macmillan.

Kopnina, H. (2012) 'Education for sustainable development (ESD): the turn away from "environment" in environmental education?', *Environmental Education Research*, vol. 18, no. 5, pp. 699–717.

Kopnina, H. and Blewitt, J. (2015) *Sustainable Business: Key Issues*, London, Routledge.

Kuhlman, T. and Farrington, J. (2010) 'What is sustainability?', *Sustainability*, vol. 2, pp. 3436–3448.

Kumar, P. (2010) *The Economics of Ecosystems and Biodiversity: Ecological and Economic Foundations*, London: Earthscan.

MacNeill, J. (2006) 'The forgotten imperative of sustainable development', *Environmental Policy and Law*, vol. 36, no. 3/4, pp. 167–170. Epigraph used with permission from IOS Press.

Mastny, L. (ed.) (2014) *State of the World 2014: Governing for Sustainability*, Washington: Island Press.

MEA (2005) Ecosystems and Human Well-being: Opportunities and Challenges for Business and Industry, Millennium Ecosystem Assessment, online, available at: www.millenniumassessment.org/documents/document.353.aspx.pdf (accessed 26 January 2013).

Meadows, D., Randers, J. and Meadows, D. (2004) *Limits to Growth: the 30-year Update*, Vermont: Chelsea Green.

Neumayer, E. (2007) 'A missed opportunity: the Stern Review on climate change fails to tackle the issue of non-substitutable loss of natural capital', *Global Environmental Change*, vol. 17, no. 3/4, pp. 297–301.

OECD (2011) Towards Green Growth, Organisation for Economic Co-operation and Development, online, available at: www.oecd.org/greengrowth/48224539.pdf (accessed 10 July 2014).

Orr, D. (2003) Four Challenges of Sustainability, online, available at: www.ratical.org/co-globalize/4CofS.pdf (accessed 19 July 2014).

Ott, K. and Doring, R. (2008) 'Strong sustainability and environmental policy: justification and implementation', in *Sustaining Life on Earth: Environmental and Human Health through Global Governance*, ed. C. Soskolne, New York: Lexington Books.

Pearce, D. (1993) *Blueprint3: Measuring Sustainable Development*, London: CSERGE and Earthscan.

Rees, W. (2008) 'Toward sustainability with justice: are human nature and history on side?', in *Sustaining Life on Earth: Environmental and Human Health through Global Governance*, ed. C. Soskolne, New York: Lexington Books.

Rees, W. (2010) 'What's blocking sustainability? Human nature, cognition and denial', *Sustainability: Science, Practice and Policy*, vol. 6, no. 2 (ejournal), online, available at: http://sspp.proquest.com/archives/vol. 6iss2/1001–012.rees.html (accessed 21 June 2014).

Rennings, K. and Wiggering, H. (1997) 'Steps towards indicators of sustainable development: linking economic and ecological concepts', *Ecological Economics* vol. 20, no. 1, pp. 25–36.

Rist, G. (1997) *The History of Development: From Western Origins to Global Faith*, London: Zed Books.

Rolston III, H. (2012) *A New Environmental Ethics: the Next Millennium for Life on Earth*, New York: Routledge.

Rostow, W. (1960) *The Stages of Economic Growth: a Non-Communist Manifesto*, Cambridge: Cambridge University Press.

Schumacher, E. (1973) *Small is Beautiful*, New York: Harper Torchbooks.

Shiva, V. (1992) 'Recovering the real meaning of sustainability', in *The Environment in Question*, eds. D. Cooper and J. Palmer, London: Routledge.

Simon, D. (2008) 'Neoliberalism, structural adjustment and poverty reduction strategies, in *The Companion to Development Studies*, eds. V. Desai and R. Potter, London: Arnold.

Simon, J. (1998) *The Ultimate Resource 2*, New Jersey: Princeton University Press.

Sloan, K., Klingenberg, B. and Rider, C. (2013) 'Towards sustainability: examining the drivers and change process within SMEs', *Journal of Management and Sustainability*, vol. 3, no. 2, pp. 19–30.

Solow, R. (1974) 'The economics of resources or the resources of economics', *American Economics Review Proceedings*, vol. 64, pp. 1–14.

Solow, R. (1993) 'Sustainability: an economists perspective', in *Economics of the Environment: Selected Readings*, eds. R. Dorfman and N. Dorfman, New York: Norton.

Soskolne, C. (2008) 'Preface' to *Sustaining Life on Earth: Environmental and Human Health through Global Governance*, ed. C. Soskolne, New York: Lexington Books.

Soulé, M. E. (2001) 'Does sustainable development help nature?', *Wild Earth*, Winter 2000/2001, pp. 56–64.

Sukhdev, P. (2013) 'Transforming the corporation into a driver of sustainability', in *State of the World 2013: Is Sustainability Still Possible?*, ed. L. Starke, Washington: Island Press.

Terborgh, J. (1999) *Requiem for Nature*, Washington: Island Books.

UNEP (2011) *Towards a Green Economy: Pathways to Sustainable Development and Poverty Eradication*, United Nations Environment Programme, online, available at: www.unep.org/greeneconomy (accessed 2 December 2014).

Victor, P. (2008) *Managing Without Growth: Slower by Design, not Disaster*, Cheltenham, UK: Edward Elgar.

Vogt, W. (1948) *Road to Survival*, New York: William Sloan.

Wackernagel, M. and Rees, W. (1996) *Our Ecological Footprint: Reducing Human Impact of the Earth*, Gabriola Island, BC, Canada: New Society Publishers.

Ward, H. (2008) 'Liberal democracy and sustainability', *Environmental Politics*, vol. 17, no. 3, pp. 386–409.

Washington, H. (2013) *Human Dependence on Nature: How to Help Solve the Environmental Crisis*, London: Earthscan.

WCED (1987) *Our Common Future*, World Commission on Environment and Development, London: Oxford University Press.

WCS (1980) *World Conservation Strategy: Living Resource Conservation for Sustainable Development*, IUCN/UNEP/WWF.
Wijkman, A. and Rockstrom, J. (2012) *Bankrupting Nature: Denying our Planetary Boundaries*, London: Routledge.
WWF (2012) Living Planet Report. World Wide Fund for Nature, online, available at: http://d2ouvy59p0dg6k.cloudfront.net/downloads/lpr_2012_summary_booklet_final.pdf (accessed 21 July 2014).

4
ECONOMIC SUSTAINABILITY
Coming to grips with endless growth

'A person who knows that enough is enough will always have enough.'
(Lao Tzu)

Perhaps no strand of sustainability needs more demystification or is as difficult to come to grips with as *economic sustainability*. I start by stating I am not an 'economist'. I am an environmental scientist from outside economics, *looking in* – and trying to make sense of it. This could be seen as a drawback. Given what I discuss below, others may consider this to actually be an advantage. The ancient Greek root of 'economy' is *oikonomos*, 'one who manages a household', derived from *oikos*, 'house' and *nemein*, 'to manage'. From this was derived *oikonomia*, or 'management of a household or family'. Given how far apart the terms have now moved, it is interesting how similar it is to *oikos logos*, the 'study of the home', which we now called 'ecology'. The Oxford English Dictionary defines economy as 'the state of a country or region in terms of the production and consumption of goods and services and the supply of money'. 'Wealth' is another key term, being a measure of the value of all assets of worth (material prosperity). Yet wealth measured in what way? Production and consumption of goods and services at what level, and based on what ideology? Related to what biophysical limits? Does 'economy' work hand in hand with 'ecology'? Human societies have always had a 'market' (originally for barter) and traded things. So all societies historically have had an 'economy'.

At its simplest then, the economy is how we organise things in our society, how we produce food and materials, distribute and trade them, and swap skills. Economics is thus the study of how humans make their living, how they satisfy their needs and desires (Common and Stagl 2005). Barter was a common means of exchange, and Polanyi (1994) argues that before the free market economy there

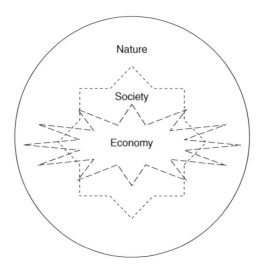

Figure 4.1 Nature, society and economy, adapted from Victor (2008) *Managing Without Growth*, figure 2.2. The economy has arms that reach into both society and Nature, but is a subset of both.

were actually three systems: redistributive, reciprocity and house-holding. Later we developed a portable means of exchange – money, the problems of which we discuss later. The economy was thus always meant to *serve* society. We shall consider whether this is still the case today. So what is it that we are 'sustaining'? From the original meaning of the term, one might think we sustain the 'home' we manage? That we would keep 'sustainable' the natural world that provides us with goods, and from which we derive services? 'Good' economics should be good management of the home, the home we study with ecology. However, this is now anything but the case. Modern economics is fraught with issues of worldview, ideology, assumptions, ignorance of ecological reality, and denial. Denying the problems of the 'growth economy' is also arguably the largest 'elephant in the room' (Zerubavel 2006) that most of us refuse to see.

Kopnina and Blewitt (2015) note that sustainability cannot be met by simply tinkering with the current economic system. To talk about 'economic sustainability' one must confront the issue that our current economy is fundamentally broken and in need of renewal (Jackson 2009, p. 15). The *growth economy* has become the 'given truth' of our times, what Ellul (1975) has called the 'chief sacred' in society. More than anything else in our society, the 'growth economy' remains taboo to question, and the silence of co-denial (Zerubavel 2006) concerning it remains almost supreme. This is true even in many academic, media and environmental circles. This fact is crucial to our times, central to whether we reach a truly sustainable future. 'Economics' is one of the most used (if not understood) words. The Oxford English Dictionary defines this as 'the branch of knowledge concerned with the production, consumption, and transfer of wealth'. I will briefly discuss classical, neo-classical and Keynesian economics below, as they are the backbone of the neo-classical economic synthesis in power. Then I will discuss the steady state economy, degrowth and the circular economy, and UNEP's 'green economy'. There are of course many other schools of economic thought, such as Marxian, Austrian, institutional, evolutionary and even complexity economics (Foxon *et al.* 2012), which will not be discussed (due to space limits). There has also been extensive discussion of the 'service' economy, where people trade services, and this will be discussed briefly.

Classical economics is regarded as the first modern school of economic thought. Its major developers include Adam Smith, Jean-Baptiste Say, David Ricardo, Thomas Malthus and John Stuart Mill. Adam Smith's (1776) 'The Wealth of Nations' is considered to mark the beginning of 'classical' economics. Classical economists argued that free markets *regulate themselves* when free of any intervention. Smith referred to a so-called 'invisible hand' that will move markets towards their natural equilibrium, without requiring any outside intervention. It was his shorthand for the ability of the free market to allocate factors of production, goods and services to their most valuable use. He argued if everybody acts from self-interest (spurred on by the profit motive), then the economy will work more efficiently and productively than it would if directed by a 'central planner'. Smith thought it was as if an 'invisible hand' guided the actions of individuals to combine for the common good. However, he also recognised it was not infallible, and that government action might be needed (such as to enforce property rights) (*Economist* 2013). He also realised it required social ethics, and in fact Smith (1759) in 'The Theory of Moral Sentiments' stated that the market was 'so dangerous that it presupposes the moral force of shared community values as its necessary restraining context'. Daly (1991, p. 13) notes that the 'invisible hand' is meant to lead private interest to serve public welfare. However, he also noted that 'while the invisible hand looks after the private sector, the invisible foot kicks the public sector to pieces'. Classical economics was followed by 'neoclassical economics' in the mid-nineteenth century.

Neoclassical economics is a term variously used for approaches to economics focusing on the determination of prices, outputs and income distributions in markets through supply and demand, often mediated through a hypothesised 'maximisation of utility' (satisfaction) by income-constrained individuals, and of profits by cost-constrained firms using 'rational choice theory' (Campus 1987). Neoclassical economics dominates microeconomics, and together with Keynesian economics forms the neoclassical 'synthesis' that dominates mainstream economics today (Clark 1998). The key assumptions of neoclassical economics are discussed later.

Keynesian economics is the view that in the short run (especially during recessions) economic output is strongly influenced by total spending in the economy. John Maynard Keynes (1936) published 'The General Theory of Employment, Interest and Money' during the Great Depression. Keynesian economists argue that private sector decisions sometimes lead to inefficient macroeconomic outcomes, which require *active policy responses* by the public sector to stabilise the economy (e.g. monetary policy actions by the Central Bank) (Sullivan and Sheffrin 2003). Keynesian economics advocates a 'mixed economy', predominantly private, but with a role for government intervention during recessions. In the developed nations, Keynesian economics served as the standard economic model during the latter part of the Great Depression, the Second World War, and the post-war economic expansion (1945–1973). It lost some influence following the stagflation of the 1970s (Fletcher 1989).

We should speak briefly of 'capitalism'. It consists of private ownership of the means of production, along with allocation and distribution provided by the market. Individual maximisation of profit by firms and the maximisation of satisfaction (utility) by consumers provides the motive force, while competition, the existence of many buyers and sellers in the market, provides the 'invisible hand' that supposedly leads private interest to serve public welfare (Daly 1991, p. 13). Since the 1980s the general embrace of neoclassical economics in society has led to increasing inequalities of wealth (see Chapter 6) and more frequent and severe booms and crashes. The ideological clash in the United States between Keynesians (generally Democrats) and Neoliberals (generally Republicans) is likely to continue. However, Heinberg (2011, p. 39) argues it will yield little light if both philosophies conceal 'the same fundamental errors: that economies can perpetually grow'. Also, the inclusion of 'land' within capital amounts to saying Nature is merely a subset of the human economy. It also suggested that natural resources could always be substituted with some other form of capital (money or technology), as in the 'weak sustainability' (Heinberg 2011, p. 40) we discussed in Chapter 3 (which denies ecological realities).

Worster (1988, pp. 11–12) gives an excellent discussion of the 'industrialism' that became the shield-maiden of capitalism:

> The capitalists ... promised that, through the technological domination of the earth, they could deliver a more fair, rational, efficient and productive life for everyone.... Their method was simply to free individual enterprise from the bonds of traditional hierarchy and community, whether the bondage derived from other humans or the earth.... That meant teaching everyone to treat the earth, as well as each other, with a frank, energetic, self-assertiveness.... People must ... think constantly in terms of making money. They must regard everything around them – the land, its natural resources, their own labor – as potential commodities that might fetch a profit in the market. They must demand the right to produce, buy, and sell those commodities without outside regulation or interference ... as wants multiplied, as markets grew more and more far-flung, the bond between humans and the rest of nature was reduced to the barest instrumentalism.

Sukhdev (2010) says the root causes of biodiversity loss lie in the nature of the human relations with Nature. He sees the root cause of the problem as being our dominant economic model, which:

> promotes and rewards more versus better consumption, private versus public wealth creation, human-made capital versus natural capital. This is the 'triple

whammy' of self-reinforcing biases that leads us to uphold and promote an economic model in which we tend to extract without fear of limits, consume without awareness of consequences and produce without responsibility for third party costs, the so-called 'externalities' of business.

There is also the key issue of *money and finance*, which have now become divorced from reality. Money is being used to create money and shuffled around in a shell game where nothing tangible is being produced (Dietz and O'Neill 2013). This has led to a major debt crisis, which could soon cause the collapse of the growth economy (Heinberg 2011). Palley (2014) argues that the current 'financialisation' means that the financial sector has become the new master of the broader economy and an engine for consuming resources. Financialisation lobbies for the deregulation of the finance market and provides credit that creates asset price bubbles, hence it has been said to be at the core of our economic difficulties (Palley 2014, p. 178). The existing monetary system is thus now inherently unsustainable. Most of the money supply comes from 'fractional reserve banking', where banks are only required to retain a fraction of the deposits received (commonly 10 per cent). The rest they loan out at interest, and loans are then deposited in other banks, which can also loan it out. If a government credits $1 million to a bank, then with a 10 per cent reserve requirement, banks can then create $9 million of 'new money', so most money today is created as interest-bearing debt. Total US debt is now $50 trillion, and our relentless pursuit of debt-driven growth has contributed to the global economic crisis (Costanza *et al.* 2013). This system is destabilising, for when the economy booms, banks loan money hugely, stimulating growth and more lending. An economic slowdown then leads to widespread loan default, and creates a self-reinforcing downward economic spiral. This system also transfers resources to the financial sector, and prioritises investment in 'market' goods over public goods (e.g. hospitals, schools, libraries). Lastly, it is ecologically unsustainable, as debt can grow exponentially, while future production confronts ecological limits and cannot possibly keep pace. Eventually, the exponentially increasing debt must exceed the value of current real and future potential wealth, and the system collapses (Costanza *et al.* 2013).

The underlying assumptions of neoclassical economics

There are assumptions that neoclassical economics makes about how the world works (Diesendorf and Hamilton 1997; Costanza *et al.*, 2013). These should (indeed must) be examined if we are to reach economic sustainability. These include:

1   Strong *anthropocentrism*. Nature is seen as 'just a resource' to be used to provide the greatest 'utility' to the greatest number of people. Rolston (2012, p. 37) notes that if neoclassical economics is the 'driver we will seek' for our society, then it will result in 'maximum harvests in a bioindustrial world', as the current economic model is extractive in nature, and commodifies the land. Land

becomes merely 'resources' and 'natural capital'. Such maximum harvests will not consider the limits (or tipping points) of ecosystems.

2 The idea that the *free market* will control all that is needed, that the 'invisible hand' will regulate things for human benefit (Daly 1991, p. 13). This is a key given truth and has become almost a religion (Daly 2008). The deification of the free market has also been shown to be an underlying ideological reason for conservatives to deny climate change (Oreskes and Conway 2010). Stiglitz (2002) argues the invisible hand is 'invisible' because 'it is not there'. Common and Stagl (2005, p. 350) note the invisible hand does not in fact work. Market failures of various kinds mean that actual market outcomes are not efficient. Rees (2010) notes that: 'Unsustainability may be the greatest example of market failure'. Renner and Prugh (2014) note that markets don't have a social conscience, environmental ethic or long-term vision, no do they conserve the 'public good' (Wijkman and Rockstrom 2012, p. 134). Achieving efficiency does not guarantee equity, between either those alive at one point in time, or at different points in time. It doesn't consider intergeneration equity. Even under 'ideal' conditions, market outcomes may be very unfair.

3 The idea that the economy can *grow forever* in terms of continually rising GDP, which increased by an astounding 25-fold over the last century (Dietz and O'Neill 2013). Daly (1991, p. 8) notes that 'economic growth is the most 'universally accepted goal' in the world and that: 'Capitalists, communists, fascists and socialists all want economic growth and strive to maximise it'.

4 The *refusal to accept any biophysical limits to growth*, for when classical economics was developed, limits were distant (Daly 1991, p. 19). However, neoclassical economics today still does not acknowledge any limits on a finite Earth, and economists are not taught how Nature actually works (Wijkman and Rockstrom 2012, p. 17). Daly (1991, p. 187) notes that three inter-related conditions: finitude, entropy and complex ecological interdependence – combine to provide the biophysical limits to growth.

5 A *circular theory of production* causing consumption that causes production in a never-ending cycle. Daly (1991, p. 184) notes that real production and consumption are in no way circular. The growth economy sees outputs returned as fresh inputs and he notes ironically this requires we 'discover the secret of perpetual motion' (ibid. p. 197). 'Money fetishism' is the idea that money flows in an isolated circle, and thus so can commodities. Daly and Cobb (1994, p. 37) argue that the neoclassical view believes that if money balances can grow bigger forever at compound interest, then so can real GNP. They respond ironically 'so can pigs and cars and haircuts'. This is a classic 'fallacy of misplaced concreteness' we discuss later.

6 Neoclassical economics *ignores the Second Law of Thermodynamics* and fails to consider 'entropy' as a key feature of economics and reality. Georgescu-Roegen (1971) and Daly (1991) explore this. Thermodynamics shows that we do not create or destroy anything in a physical sense, we merely transform or rearrange it. The inevitable cost of arranging greater order in one part of the system (the

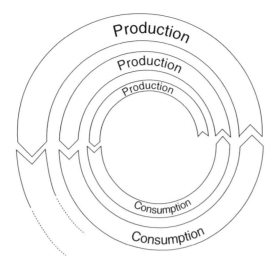

FIGURE 4.2 The assumed 'circular flow' of production and consumption in the neoclassical economy (source: adapted from figure 7 in Daly (1991) *Steady State Economics*).

human economy) is to create disorder elsewhere – Nature (Daly 1991, p. 24). 'Entropy' is a measure of the disorder in a closed system. In thermodynamics, low entropy quantities (usable energy, raw materials) move to high entropy quantities (waste heat and wastes). Entropy is the basic physical coordinate of 'scarcity'. Were it not for entropy, we could burn the same gallon of petrol over and over, and our capital stock would never wear out. Technology cannot rise above the laws of physics, so there is no question of 'inventing a way to recycle energy', even if some economists seem to imply this (Daly 1991, p. 36). Economic modelling cannot ignore entropy, otherwise it becomes irrelevant to reality. Such physical laws will not go away, and Einstein thought the entropy law the 'least likely to ever be overthrown' (Daly 1991, p. 36).

7   Environmental damage is *merely an 'externality'*. The spillover effects of market transactions have been named 'externalities'. Externalities are costs or benefits arising from an economic activity that affect somebody other than the people engaged in it, and are not reflected fully in prices. Environmental damage is known as a 'negative externality', something external to the economic model. An externality (to neoclassical economists) is seen as being worth only peripheral attention (Daly and Cobb 1994, p. 53). This is a key part of what has led to the environmental crisis. Of course, environmental crises can still occur even where externalities *are* 'internalised' (incorporated into market accounting), and Foxon *et al.* (2012) point out that this approach is inadequate for solving climate change and biodiversity loss. However, internalising such costs is a better approach than ignoring them. UNEP is seeking to cost ecosystem services and seek that they are built into economic decisions (Kumar 2010).

8   All forms of *capital can be substituted*, thus human capital can be substituted for natural capital (weak sustainability).

The above assumptions have been detailed individually by others, principally Herman Daly (Daly 1991, 1996; Daly and Cobb 1994), though rarely brought together as shown here. These assumptions show the fundamental challenge we face to reach any conception of 'economic sustainability' that functions within ecological and social realities. Looking at them from the viewpoint of environmental science, the above assumptions are bizarre and unsustainable. However, they underpin the reigning neoclassical economic synthesis to this day. The above assumptions deserved to be challenged, and some economists and scientists have been doing so for over 40 years.

The steady state economy

One of the first people to challenge the growth assumptions of neoclassical economics was economist Nicholas Georgescu-Roegen, who in 1971 wrote 'The Entropy Law and the Economic Process'. This used thermodynamics to show the absurdity of growthism. He was followed by his student, Herman Daly, an economist who coined the term the 'steady state' economy in 1973. Several others have enlarged on what has been called 'ecological economics', an economics that acknowledges the ecological limits of the planet, which considers interactions between economic systems and ecological systems (Common and Stagl 2005). Faber (2008) states that ecological economics is defined by its focus on Nature, justice and time. Issues of intergenerational equity, irreversibility of environmental change, uncertainty of long-term outcomes, and sustainability guide ecological economics. The key points of the steady state economy (Daly 1991, p. 7) are:

1 Constant population (at an ecologically sustainable level), and
2 A constant low level of *throughput* in materials and energy.

'Throughput' is the entropic physical flow of matter-energy from Nature's source through the human economy and back to Nature's sinks. Neither population nor artefacts (made things) can continue to grow forever. What is held constant is capital stock in the broad sense: capital goods, consumer goods and human population. What is *not* held constant is growth in our culture, knowledge, goodness and ethics. Czech (2000, 2013), Victor (2008), Jackson (2009), Simms et al. (2010), Heinberg (2011) and Dietz and O'Neill (2013) have continued to develop this theme.

The steady state economy is deduced from first principles regarding physical laws and ecological limits. If the world is a finite, complex system that evolved using a fixed rate of flow of solar energy, then any economy that seeks the indefinite expansion of its stocks and energy use will sooner or later hit limits. This is logically trivial, a truism, but it is not trivial psychologically or politically (Daly 1991, p. 47). If humanity's behaviour should be governed by values of 'enoughness', stewardship, humility and holism, then it follows that attitudes of 'more forever', technical arrogance and aggressive analytical reductionism should be rejected (Daly 1991, p. 48).

The vision of neoclassical economics is that the economy is an isolated system in which exchange value circulates between firms and households. Daly (1991, p. xiii) argues this is like saying that nothing enters from the environment, nothing exists in the environment, nothing exits to the environment. Connections to the larger environment cannot be 'abstracted' from, without losing the most essential fact, as neoclassical economics has done. This contrasts to reality, where a linear flow of energy and materials moves from low entropy (usable energy, rich resources) to high entropy (heat and waste) (Daly 1991, p. 184). In neoclassical economics, it doesn't matter how big the economy is relative to the environment or if it impacts disastrously. For the steady state economy however, the vision is that the economy is an open subsystem of a finite and non-growing ecosystem. The economy lives by importing low-entropy matter-energy (raw materials) and exporting high entropy matter-energy (waste).

Degrowth and the circular economy

We should also discuss briefly two other key terms. Given that Western economies have already grown too large (in terms of being ecologically sustainable), some scholars point out we also need 'degrowth' (Latouche 2010; Assadourian 2012; Dietz and O'Neill 2013). The West, it is argued, should contract its economy, so that the developing world can expand somewhat to alleviate poverty, with the resulting steady state economy being of smaller size and within the Earth's limits. This idea in regard to carbon emissions has been called 'contraction and convergence' (Meyer 2000). Degrowth acknowledges that stopping growth may not be enough when Western economies are already too large. Some scholars however argue that degrowth would be 'unstable' (e.g. Jackson 2009).

There has been much recent discussion of a 'circular' economy (e.g. Wijkman and Rockstrom 2012; EMF 2012). This is essentially the application of the 'cradle to cradle' philosophy (Braungart and McDonough 2008) to the economy. That means that products are eco-designed and eco-effective (see Chapter 7). It means that as little energy and materials are used as possible, and ties in with terms such as the 'blue economy' (Pauli 2010) and 'performance economy' (Stahel 2010). The most radical aspects of the circular economy are to replace materials with 'services' and to design products to be long-lasting. Hence instead of selling things you would *lease them* and provide exceptional service. The producer would thus take responsibility for the product over its whole life cycle, and replace it when it wears out, making sure its materials are reused or recycled. In a circular economy there would be no reprehensible 'planned obsolescence' strategy (currently rampant in business). However, while it has been described as 'revolutionary' (Wijkman and Rockstrom 2012, p. 168), the circular economy is not necessarily a steady state economy (it does not discuss the population aspect). Rather, economic growth in a circular economy would not occur via a massive extraction of resources but through providing services and conserving materials. However, the 'cradle to cradle' strategies proposed under a circular economy also make excellent sense as part of a steady state economy.

Endless growth and 'growthmania'

Rees (2010) notes that the 'flawed assumption that human well-being derives from perpetual growth' has distorted the lives of more people than any other cultural narrative in history, and is 'lodged in the heart of the (un)sustainability conundrum'. Meadows *et al.* (2004, p. 12) note that the global economy is already so far above sustainable levels that there is very little time left for the fantasy of an infinite globe. For most of our history, we managed without growth, the last few centuries being the exception (Victor 2008, p. 24). John Stuart Mill (1859) thought a stationary state of capital was a 'considerable improvement on our present condition'. Economists such as Mishan (1967), Schumacher (1973), Zolotas (1981), Daly (1991, 1996), Douthwaite (1999), Booth (2004), Porritt (2005), Siegel (2006), and Jackson (2009) question the desirability and feasibility of continued economic growth (based on increasing population and resource use). The book 'Limits to Growth' (Meadows *et al.* 1972) questioned whether growth could continue forever in a finite world. Lowe (2005, p. 8) notes it was fiercely attacked because it challenged the fundamental myth of modern society: unlimited growth. Hubbert (1993) explains that during the last two centuries we have known nothing but exponential growth, and so have evolved an 'exponential growth culture' dependent on the continuance of exponential growth for its stability. This culture is 'incapable of reckoning with problems of non-growth' (Daly and Cobb 1994, p. 408). Daly (in Lowe 2005) says most economists 'chase an assumption of wants along the road of infinite growth'. Daly (1991, p. 183) argues that economic growth is unrealistically held to be 'the cure for poverty, unemployment, debt repayment, inflation, balance of payment deficits, the population explosion, crime, divorce and drug addiction'.

Economic growth is thus seen as the panacea for everything. Sometimes commitment to growth may be promoted in the guise of 'free trade', 'competitiveness', 'productivity' – or even as 'sustainable development' (Victor 2008, p. 15). People cling to growth because they so badly need some form of 'hope' (Meadows *et al.* 2004, p. 261). Perhaps nothing stands in the way of sustainability so much as the notion that we can *spend* our way out of unsustainability (Czech 2013, p. 195). World leaders seek growth above all else. This fixed idea is based on false premises, and an economic theory that ignores limits and ecological reality. Daly (1991, p. 99) notes that the verb 'to grow' has become *twisted*. We have forgotten its original meaning: to spring up and 'develop to maturity'. The original notion included maturity or 'sufficiency', beyond which accumulation gives way to maintenance. Thus growth gives way to maturity, a *steady state*. Growth is not synonymous with 'betterment'. To grow beyond a certain point can be disastrous. Edward Abbey (1977) observed that 'endless growth is the ideology of the cancer cell'. Cafaro

(2012, p. 314) concludes that modern economic theory 'reads as if cancer had found a voice'. 'Growthmania' is not counting the costs of growth. Society today takes the real costs of increasing GNP (as measured by expenditures incurred to protect ourselves from the unwanted side effects of production) and adds these expenditures to GNP, rather than subtract them. We thus count real costs as benefits, and this is 'hypergrowthmania' (Daly 1991, p. 99). Economic growth is cannibalistic, it feeds on both Nature and communities and shifts unpaid costs back onto them as well. The shiny side of development is often accompanied by a dark side of displacement and dispossession, this is why economic growth has time and again produced impoverishment next to enrichment (Sachs 2013, p. 125).

One fascinating historical note is that none of the key classical and neoclassical economists (such as Smith, Mill and Keynes) thought an economy *could* grow forever (Dietz and O'Neill 2013). They all spoke of a growth period, after which the economy would level off. Mill (1859) thought a stationary state of capital was a 'considerable improvement on our present condition' (Daly 1991, p. 14). Once we have gone beyond the optimum, and marginal costs exceed marginal benefits, growth will make us worse off. We have then reached *uneconomic growth* (Daly 2008). However, our experience of diminished well-being will be blamed on 'product scarcity'. The orthodox neoclassical response will then be to advocate *increased growth* to fix this. In the real world of ecological limits, this will make us even less well off, and will lead to advocacy of 'even more growth'. As Daly (1991, p. 101) notes: 'The faster we run, the behinder we get'. Daly argues that environment degradation today is largely a disease induced by economic physicians who treat the sickness of unlimited wants by prescribing unlimited production. But one does not cure a treatment-induced disease by increasing the treatment dosage. Daly (1991, p. 101) quotes Luke (in the Bible): 'Physician, heal thyself!' Healing our economy requires accepting the reality that the economy cannot grow forever.

It has been argued that there are three types of growth – physical growth (cannot grow forever); economic growth (e.g. money flows, incomes, which has no theoretical limit); and human welfare growth (Ekins 2009). However, the lack of a 'theoretical limit' to economic growth may be because it ignores ecological limits. Ecological economists speak of growth as 'quantitative physical increase' in the matter–energy throughput (Daly 2013). However, the 'economic growth' Ekins describes above, if it has no physical impacts, clearly isn't based on increasing population or throughput of energy and resources. 'Economic development' is a term applied to changes in an economy that do not involve population increase or increasing throughput (Czech 2013). GDP may still rise from economic development, but it is not due to increasing population or throughput of resources.

'Decoupling' is a much-discussed term. If we grow our economy by becoming *cleverer* in the way we do things, then this is fine in an SSE, which states that growth should not occur by population increase or increased throughput. To what extent can this proceed? A 'service economy' is often portrayed as decoupling economic growth from impact. However, any growth in population or artefacts (houses, infrastructure) must increase environmental impact. UNEP (2011a) in its 'green economy' speaks of completely decoupling economic growth from impacts. The question remains as to 'just how far' can we *actually* decouple throughput of resources and live a similar life? Factor 5 (von Weizsäcker et al. 2009) argues we could reduce energy and material use by 80 per cent. This is a worthy goal, one for which the technologies already exist (Wijkman and Rockstrom 2012, p. 75) but how successful have we been in reaching it?

The scope for de-coupling growth in production and consumption from environmental degradation is limited and the decoupling is unable to keep up with unlimited growth (Næss 2011). Victor (2008) notes decoupling slows down the rate at which things get worse, but *does not turn them around*. He notes further that some modest decoupling of material flows occurred in some industrialised countries from the mid-1970s to the mid-1990s, but total material throughput still increased. Despite increases in efficiency, decoupling GDP and throughput has yet to manifest itself as an increase in GDP combined with a decline in throughput (Victor 2008). Similarly, Matthews et al. (2000) found no evidence that moderate decoupling led to absolute reduction in resource throughput and Wijkman and Rockstrom (2012, p. 152) concluded the same. So, to date, it seems impossible to *fully* decouple economic growth from physical impact. Some forms of positive increase in GDP that don't rely on population and increased throughput may continue under an SSE (though these should more properly be called 'economic development'). However, talk of '100 per cent decoupling' may just be wistful thinking that allows 'business-as-usual' growth to continue. Daly (1991) believes it was an illusion to think that growth could continue by becoming ever less materially intensive and service-oriented, and Czech (2013) concurs. Welzer (2011) concludes that the decoupling debate maintains the illusion that we can 'just make minor adjustments'.

Regarding growth, Victor (2008, p. 170) concludes:

> At some point, fundamental questions of growth for what, for whom and with what consequences will be asked by more and more people until there is a shift in societal values away from a growth-first policy. Some glimmers of that shift are discernible today.

Reductionism, economic modelling and the 'fallacy of misplaced concreteness'

Reductionism is the idea that you could understand the world by examining smaller and smaller pieces of it, which science and academia mostly do. However, the risk is that you lose overall perspective, become so focused on a particular part you fail to see the whole, as has happened with neoclassical economics. A key problem is the 'fallacy of misplaced concreteness', which Georgescu-Roegen (1971) thought was the key sin of standard economics. Philosopher Alfred Whitehead (1929) defines this as: 'neglecting the degree of abstraction involved when an actual entity is considered, merely so far as it exemplified certain categories of thought'. In other words, reductionism and modelling pose the risk that the model becomes 'more real' than reality. The abstractions necessary to create mechanistic models 'always do violence to reality' (Daly 1991, p. 46). Policy decisions are 'determined by mathematical theorems whose virtue is their deductive fruitfulness rather than their connection to the real world'. The abstraction has gone too far and the practitioners are not aware of this, for 'the fallacy of misplaced concreteness is too pervasive' (Daly and Cobb 1994, p. 96). Those who habitually think in terms of abstract reductionist models are especially prone to this fallacy. Mathematical models and computer simulations are the hallmark of neoclassical economics. Such elaborate logical structures prize theory over fact, and reinterpret fact to fit theory (Daly and Cobb 1994, p. 38).

'Resources are infinite', techno-centrism and substitution

Daly (1991, p. 108) points out that another assumption of neoclassical economics is that 'resources are infinite', that when one dries up there will 'always be another', and technology will always find cheap ways to exploit this (techno-centrism). As long as neoclassical economics continues to ignore the absolute dimension of scarcity, then it has 'uncoupled itself from the world and become irrelevant' (Daly 1991, p. 108). Mining of ores and minerals grew 27-fold in the twenty-first century (UNEP 2011b). Some economists have denied the relevance of matter and energy. For example, Gilder (1981) spoke of 'overcoming the materialistic fallacy', the illusion that resources and capital are things that can 'run out', rather than 'products of the human will and imagination which in freedom are inexhaustible'. Most economists *do* actually recognise that production and consumption have physical dimensions, but most also have largely remained silent about stating this (Daly 1991, p. 198).

Daly (1991, p. 283) shows an important example of misplaced concreteness is the neoclassical proposition that human-made capital is a 'near perfect substitute' for natural resources (Nordhaus and Tobin 1973; Stiglitz 1979). There is the absurd implication that 'we could build the same house with one tenth of the lumber if only we had more saws or hammers'. Yet this substitutability argument is often cited in order to argue that natural resource scarcity need not limit growth. Economist Julian Simon (1981) has stated about substitution 'you see, in the end copper and oil come out of our minds. That's really where they are'. Such views are still influential, and are rarely criticised by mainstream economists (Daly and Cobb 1994, p. 109). There is thus a widespread modern neoclassical economic view that human capital can substitute for land or Nature (weak sustainability), and hence the goal of increasing capital can proceed without attention to what is *physically happening* to the land (ibid., p. 110). This is an important driver of the environmental crisis.

Classical (even neoclassical) economic theory developed at a time when the environment was considered an 'infinite resource', but the scale of our economy is now vastly larger (Daly 1991, p. 185). The concept of *throughput* should now displace the 'circular flow' model in economic theory. Daly warns us not to get carried away with the human mind being an 'ultimate resource' that can generate endless growth, for the mind is not independent of the body. Before leaving resource use, we should discuss Zeno's Paradox of 'Achilles and the Tortoise'. Achilles gives the tortoise a head start, but whenever Achilles reaches somewhere the tortoise has been, he still has farther to go. Because there are an infinite number of points that Achilles must reach (where the tortoise has already been) – he can never overtake the tortoise! The paradox actually confuses an infinity of subdivisions of a finite distance with an infinity of distance. Yet this is what Simon (1981) similarly does with resources. He confuses an infinity of possible boundary lines between 'copper' and 'non-copper' with an infinite amount of copper. There is (after all) only a finite amount of copper on Earth. The flawed logic of Zeno's paradox however is still commonly argued to suggest that resources have no limits (Daly and Cobb 1994, p. 263).

The ethics of economics

Is there an ethics of economics? The two terms are rarely spoken of together. However, if we are to reach 'economic sustainability', they should be. Robertson (1954) once asked 'What does the economist economise?'. His answer was 'love'. Daly (1991, p. 4) says an economist is similar to how Oscar Wilde described a cynic: 'A man who knows the price of everything and the value of nothing'. Bauman and Rose (2009) note that modern economists (and economics students) are less generous and more selfish than others. Unless the underlying growth paradigm and its supporting values are altered, 'all the technical prowess and manipulative cleverness in the world' will not solve our problems, and in fact will make them worse (Daly 1991, p. 2). Daly (2008) says: 'the neoclassical view is that man will surpass all limits and remake Creation to suit his subjective preferences, which are considered the root of all value. In the end, economics is religion'. Daly and Cobb (1994, p. 21) note: 'Before this generation is the way of life and the way of death'. They conclude that at a 'deep level of being', they:

> find it hard to suppress a cry of anguish, a scream of horror. We humans are being led to a dead end, we are living by an ideology of death and accordingly we are destroying our own humanity and killing the planet.

Originally, economics started as a branch of 'moral philosophy', and ethics was at least as important as the analytic content. However, economic theory has become more and more top heavy, with layer upon layer of abstruse mathematical modelling, erected above the shallow concrete foundation of fact (Daly 1991, p. 3). Daly (1991, p. 20) notes that the temper of the modern age resists any discussion of the 'ultimate end'. *Why are we doing what we do?* Teleology and 'purpose', the dominant concepts of an earlier age, have apparently been banished. Economics followed along by reducing ethics to the level of 'personal tastes'. Individuals set their own priorities, and economics became simply the 'mechanics of utility and self-interest'. It thus divorced itself from ethics. To do 'more efficiently' that which should 'not be done in the first place' is no cause for rejoicing, Daly concludes. The big problems of overpopulation and overconsumption 'have no technical fixes but only difficult moral solutions' (Daly 1991, p. 39). The steady state economy thus threatens the Faustian covenant with 'Big Science' and high technology, for it argues all things in fact are *not possible* through technology. For these reasons the steady state economy is resisted by orthodox economists (Daly 1991, p. 39). It is also resisted by techno-centrists and Cornucopians. The ethical dimensions of dealing with the growth economy are thus enormous (see Chapter 8).

Daly (1991, p. 248) suggests that society *could* accept the eventual destruction of life-support capacity as the price we must pay for 'freedom from restriction of individual rights to grow'. However, he observes: 'It is widely believed by persons of diverse religions that there is something fundamentally wrong in treating the Earth as if it were a business in liquidation'. Daly and Cobb (1994, p. 379) conclude that a 'sustained willingness to change depends on a love of the earth that humans once felt strongly but that has been thinned and demeaned as the land was commodified'. Hence we cannot afford to let economics remain an *ethics-free zone*. The current corporate ethic seems to be to 'use resources as fast as possible until they're gone' (Heinberg 2011, p. 254). Clearly, a first step is to assert that economics *must* have an ethics, and that the current ethic is wrong and unsustainable. We discuss ethics further in Chapter 8.

The 'green' economy

There has been recent discussion of a 'green economy', especially in regard to the UN Rio+20 Summit in 2012. Interestingly, this has been almost without any reference to the steady state economy or ecological economics. UNEP (2011a) define the green economy as: 'one that results in improved human well-being and social equity, while significantly reducing environmental risks and ecological scarcities ... one which is low carbon, resource efficient and socially inclusive'. UNEP (2011a, p. 1) says it does not replace 'sustainable development', but that there is now a 'growing recognition that achieving sustainability rests almost entirely on getting the economy right'. UNEP argues that there is a widespread myth that 'there is an inescapable trade-off between environmental sustainability and economic progress. There is now substantial evidence that the "greening" of economies neither inhibits wealth creation nor employment opportunities'.

UNEP's green economy is still very much a growth economy (though it mostly calls this 'progress'), and UNEP states 'the greening of economies is not generally a drag on growth but rather 'a new engine of growth'. At the same time it acknowledges that the 'economic growth of recent decades has been accomplished mainly through drawing down natural resources ... allowing widespread ecosystem degradation and loss'. The report has three key findings. First, 'greening' not only generates increases in wealth (in particular a gain in natural capital), but also produces a higher rate of GDP growth. Second, there was an inextricable link between poverty eradication and better conservation of Nature. Third, in a transition to a green economy, new jobs are created, which over time exceed the losses in 'brown economy' jobs. However, there is a period of job losses in transition, which requires investment in re-skilling and re-education. UNEP (2011a, p. 22) conclude that moving towards a green economy has the potential to achieve sustainable development and poverty eradication on a scale and speed 'not seen before'.

The need to stabilise population is not discussed in the report. Resource use is discussed, and a central challenge is seen to be to 'decouple growth absolutely from material and energy intensity'. However, almost all economic production requires the transformation of raw materials (Costanza et al. 2013), so it is highly unlikely that economic growth could be *absolutely decoupled*. The green economy thus seems to ignore the laws of physics and ecological reality. It does not break free of the assumptions of neoclassical economics, and indeed portrays itself as a 'new engine of growth'. It makes advances towards being low carbon and low resource use, but falls into denial about the underlying problems of endless growth on a finite planet. 'World in Transition' (WBGU 2011) is an excellent German blueprint for sustainability, with useful strategies to decarbonise and dematerialise the economy. However, it too assumes a growth economy, with an economic growth rate of 2–3 per cent. In summary, so far, attempts to define a 'green economy' have been unable to kick the growth habit. UNEP's green economy does some necessary things (low carbon and material use) but is not sufficient to achieve sustainable human well-being (Costanza et al. 2013).

What *should* economic sustainability mean?

Nothing needs to be demystified so much as 'economic sustainability'. So how does economics come into sustainable development? Victor (2008, p. 19) argues that to date the commitment to 'sustainable development' is best understood as 'more of the same' rather than a radical departure from economic growth as the top policy objective. It has been argued by Sukhdev (2013, p. 143) that there is emerging consensus among governments and business leaders that 'all is not well with the market-centric economic model' that dominates today. Similarly, Alperovitz (2014, p. 191) observes that the current political and economic model is incapable of addressing the big sustainability challenges we face. Sukhdev believes the 'finger of blame' should be pointed at the corporate world, which must be brought to the table as 'planetary stewards' rather than value-neutral agents that are 'free-riding their way to global resource depletion'.

Economic sustainability in a finite world cannot be about endless economic growth. Future human well-being depends on slowing economic growth, not accelerating it (Staples and Cafaro, 2012, p. 294). Economic sustainability has to be about creating an economy that is sustainable over the truly long-term. This means not damaging the ecosystem services that underpin our society. Economic sustainability thus cannot mean continuing 'business-as-usual' along the neoclassical model. It requires returning the economy to being a 'servant' of society, not its master. It means questioning and abandoning most of the assumptions that underlie the neoclassical economic synthesis in control today. It means discarding our addiction to 'evermoreism' (Boyden 2004). That means moving to a steady state economy, and in fact means 'degrowth' in the developed world, with some further growth in the developing world (Daly 2012), where the final overall per capital resource use for everyone is *lower*. This might be at a level similar to what Australia had around 1960 (Lowe 2005) or 1970 (Turner 2011). This is not such a terrifying prospect, and does not mean going back to 'live in caves'. As Gandhi noted, there is enough for everyone's need but not for everyone's greed (Pyarelal 1958).

One of the key arguments against a steady state economy is usually that we just *have* to keep growth to 'create jobs'. This was not always seen as true in economics. Domar noted that there was hardly a trace of interest in economic growth as a policy objective in the official or professional literature of Western countries *before 1950* (quoted in Arndt 1978). There is in fact no 'given truth' that we must have growth to have jobs. Nor should there be, for rapid growth economies have *not* in fact brought full employment. For example, there were more Canadians with incomes less than the 'Low Income Cut Off' in 2005 than in 1980, despite real Canadian GDP having grown by 99.5 per cent (Victor 2008). Economic growth in Canada since 1980 has not eliminated unemployment or poverty, the distributions of income and wealth have become more unequal. Growth has also exacerbated environment problems (Victor 2008). So can we move to an SSE and still have jobs? Is it possible to do so and even reduce unemployment? There is some modelling work that suggests this is quite possible. Victor (2008) notes it is possible to develop scenarios for a 30-year time horizon for Canada where full employment prevails, poverty is eliminated, people have more leisure, greenhouse gases are drastically reduced, and the level of government indebtedness declines in the context of a low (and ultimately no) economic growth. He uses 'econometric' models to show how this could happen.

A new model of the economy would clearly be based on the goal of sustainable well-being. It would use measures of progress such as the GPI not the GDP (see later). It would acknowledge the importance of ecological sustainability, social fairness, and real economic efficiency (Costanza *et al.* 2013). Such an efficiency would have to be way beyond the usual eco-efficiency, and instead would have to involve 'eco-effectiveness' and a radical restructure of how we make things (Kopnina and Blewitt 2015). Can we have a global economy that is not growing in material terms, but that is sustainable and provides a high quality of life for people? Costanza *et al.* (2013) argue the answer is 'yes', and list examples from past societies and current initiatives (e.g. Transition Towns, Global Eco Village Network). Integrated modelling studies, such as World3, GUMBO and LowGrow (a model of the Canadian economy that shows that a non-growing economy can be stable, Victor 2008), also suggest economic sustainability *is* achievable (Costanza *et al.* 2013). Some nations are moving in the right direction – Sweden, Denmark, Japan and Germany have arguably reached a situation in which they do not depend on high rates of growth to provide for their people (Heinberg 2011, p. 251). The idea that we *can* change our economic system to ecological economics and a steady state economy is not a 'utopian fantasy'. On the contrary, it is the neoclassical 'business-as-usual' that is the true fantasy (Costanza *et al.* 2013).

Many people agree that on a finite planet endless growth is impossible. However, they don't know 'what to do', they have lived all their lives in a growth economy and find it hard to see a 'viable path' to a steady state economy. They fear it equates to a failed growth economy, though Daly (2008) points out they are as different as night and day. Moving to a steady state economy is actually the way to *avoid* a failed growth economy and another Great Depression. However, tackling the two key issues we deny – overpopulation and overconsumption – is a daunting task, and many feel overwhelmed (see Chapter 7). However, there are many things we can do *immediately* (and comparatively easily) to move towards true economic sustainability. These are:

- A first step is to move (over two decades) to a low carbon and material use economy, as recommended by UNEP (2011a) and the WBGU (2011). This would be through appropriate technologies such as renewable energy, energy conservation (REN21 2013) and sustainable building (Godfaurd *et al.* 2005). Various analyses have shown this is perfectly feasible and economic (e.g. Diesendorf 2014).
- *Tax-shifting*, by taxing the 'bads' that degrade ecosystem services. This includes carbon pricing as a key process to control climate change, but a landfill tax has also been proposed (Brown 2011). Taxes are an effective tool for internalising negative externalities into market prices and for improving income distribution (Costanza *et al.* 2013). Way back in 1997, 2,500 economists (including six Nobel Prize winners) endorsed the concept of tax shifts for climate change (Klugman 1997), so the recognition of the need for this is not new.
- *Subsidy-shifting*, especially taking the $10 billion subsidies in Australia for fossil fuels (Elliston *et al.* 2013) and transferring them to the renewable energy industry, or the $700 billion worldwide given to damaging activities (Brown 2011).
- *Control of resource use*, both non-renewable and renewable. For non-renewable resources a depletion quota has been suggested (Daly 1991, p. 191) or a 'severance tax' at the mine-mouth or well-head (Daly 2008). Severance taxes can be revenue neutral, by phasing them in while phasing out regressive payroll or sales taxes (Costanza *et al.* 2013). For renewable resources, proper holistic pricing of ecosystem services will also reduce overuse (Kumar 2010).
- *Dematerialisation* of the economy, the highest possible decoupling of the economy from resource use and an eco-effectiveness approach (Braungart and McDonough 2008). The developed countries should aim for a goal of Factor 5 (use only 20 per cent of current energy and resources, von Weizsäcker *et al.* 2009).

- Restructure the financial system to better serve the *needs* (not wants created by advertising) of society. A Green Investment Bank and Housing Bank are useful steps (NEF 2009), as are secure national or community-based 'green bonds' (Costanza et al. 2013).
- *Cooperatives*, 'not-for-profit' corporations, and credit unions as alternatives to 'profit above all else' corporations (Heinberg 2011, p. 253). Examples are Mondragon in Spain (which employs 83,000 people), community interest companies (the United Kingdom has 6,000, Dietz and O'Neill 2013), community development corporations (Alperovitz 2014, p. 196) and 'triple-bottom-line' businesses such as benefit corporations (Cordes 2014, p. 205).
- Banks should be required to move (possibly gradually) to a *100 per cent reserve requirement*, and make their money by financial intermediation and service charges, rather than lending at interest money they 'create out of nothing' (Daly 2008). They would thus not be able to 'create' new money. Banks could lend 'time deposits' where the depositor does not have access during that time (Costanza et al. 2013).
- Create a national green energy and *transport infrastructure* (NEF 2009).
- Regain control of the finance sector, remove its political power, and ensure it serves the real world economy (Palley 2014).
- *Overseas aid* by the developed world is needed for the developing world, but this must be specifically targeted to ecologically sustainable transition projects to a steady state economy.
- A *Green GNP indicator* used by all. A number have been developed (e.g. ISEW, GPI). The Genuine Progress Indicator (GPI) peaked in 1975 and has been flat or slowly decreasing ever since, due to environmental decline (Costanza et al. 2013). A suitable indicator or indicators need to be agreed and widely applied. At the very least we need two accounts, one that measures the 'benefits' of physical growth in scale and one that measures the 'costs' of that growth (Daly 2008). Happiness (e.g. the Happy Planet Index) should become a key goal of policy (Layard 2005).
- A *tax on financial transactions*, also known as a Tobin tax where Tobin suggested it be 0.5 per cent (Daly 2008). This will deter rapid speculative finance transfers that are part of 'extreme money' (Das 2011) and which exacerbate the debt crisis.
- A *tax on advertising* (Daly 2008) as well as a ban on outdoor advertising such as Sao Paulo introduced in 2007 (Sukhdev 2013, p. 149).
- Limits on *income inequality* (see Chapter 6). This means you establish both the minimum and maximum incomes in society. Daly (2008) notes that universities and the military manage with a factor of 10–20 as the upper limit. Czech (2013) suggests 15 times as an upper limit, while the Mondragon cooperative has a maximum pay of nine times the minimum (Dietz and O'Neill 2013).
- More *flexible workdays* and 'working from home' policies (Heinberg 2011, p. 253) and a 'guaranteed jobs' policy (Dietz and O'Neill 2013).

- A *sustainability tariff structure* so that 'sustainable' countries are not disadvantaged (Heinberg 2011, p. 251) when they trade with those that are not sustainable. This questions the 'free trade' mantra, which Daly (2008) points out assumes growth is always good for all.
- *Reform of corporate law* to support sustainability (Heinberg 2011, p. 253). Corporations are responsible for 60 per cent of global GDP, but damage the environment and influence policy against sustainability (Prugh 2013, p. 112). They should cease to be a 'person' legally and should not be able to make political donations. Sukhdev (2013) suggests ways to move to a new corporate model for sustainability (see Chapter 7).
- Education on the *downsides* of the growth economy and the upsides of a steady state economy, about why sharing and sufficiency are better than shopping and insatiability (Dietz and O'Neill 2013).
- Turning our economy from an ethics-free zone to one that is based on an *Earth ethics* (Rolston 2012).
- *Green investment* (e.g. renewable energy) and *green jobs*, which together can cushion the social impacts of moving to a steady state economy (Daly 2008).

However, while the above measures are key parts of economic sustainability, we must not slip into denial about the underlying problems. Despite all the above, we need to tackle 'overpopulation' and 'overconsumption' (see Chapter 7). Economics must serve society and accept limits and ecological realities. An ecologically sustainable biosphere has to be ranked higher than an endlessly increasing GDP. True economic sustainability will live within limits. It will not mean an ever-growing economy that relies on population growth and increasing throughput of resources. Now many people might argue that the steady state economy is 'politically impossible'. Arguments for change will 'provoke strident resistance from a world "socially engineered" to worship the market god' (Rees 2010). It is true that a steady state economy faces major resistance, but increasingly, viable alternatives are being presented. There *is* another way, there has to be. It is the 'Great Work' of 'true economic sustainability' to assist the transformation, where the politically impossible will become the politically inevitable (Dietz and O'Neill 2013). Daly (2008) concludes that the 'political impossibility' of a steady state economy may be less impossible than it previously appeared. If we succeed in demystifying 'economic sustainability', then the economy, 'the management of the home', *will* protect 'ecology', the 'study of the home'. As it should and must.

References

Abbey, E. (1977) *The Journey Home*, New York: Dutton.
Alperovitz, G. (2014) 'The political-economic foundations of a sustainable system', in *State of the World 2014: Governing for Sustainability*, ed. L. Mastny, Washington: Island Press.
Arndt, H. (1978) *The Rise and Fall of Economic Growth*, Sydney: Longman Cheshire.

Assadourian, E. (2012) 'The path to degrowth in overdeveloped countries' in *State of the World 2012: Moving Toward Sustainable Prosperity*, ed. L. Starke, L., Washington: Island Press.
Bauman, Y. and Rose, E. (2009) 'Why are economics students more selfish than the rest?', Institute for the Study of Labor, Discussion Paper No. 4625, online, available at: http://ftp.iza.org/dp4625.pdf (accessed 30 September 2013).
Booth, D. (2004) *Hooked on Growth: Economic Addictions and the Environment*, Lanham, MD: Rowman and Littlefield.
Boyden, S. (2004) *The Biology of Civilisation: Understanding Human Culture as a Force in Nature*, Sydney: UNSW Press.
Braungart, M. and McDonough, W. (2008) *Cradle to Cradle: Remaking the Way we Make Things*, London: Vintage Books.
Brown, L. (2011) *World on the Edge: How to Prevent Environmental and Economic Collapse*, New York: W. W. Norton and Co.
Cafaro, P. (2012) 'Epilogue: is humanity a cancer on the Earth?', in *Life on the Brink: Environmentalists Confront Overpopulation*, eds. P. Cafaro and E. Crist, Georgia, US: University of Georgia Press, pp. 315–318.
Campus, A. (1987) 'Marginal economics', *The New Palgrave: A Dictionary of Economics*, vol. 3, p. 323.
Clark, B. (1998) *Principles of Political Economy: A Comparative Approach*, Westport, CT: Praeger.
Common, M. and Stagl, S. (2005) *Ecological Economics: An Introduction*, Cambridge: Cambridge University Press.
Cordes, C. (2014) 'The rise of triple-bottom-line businesses', in *State of the World 2014: Governing for Sustainability*, ed. L. Mastny, Washington: Island Press.
Costanza, R., Alperovitz, G., Daly, H., Farley, J., Franco, C., Jackson, T., Kubiszewski, I., Schor, J. and Victor, P. (2013) 'Building a sustainable and desirable economy-in-society-in-nature', in *State of the World 2013: Is Sustainability Still Possible?*, ed. L. Starke, Washington: Island Press.
Czech, B. (2000) *Shoveling Fuel for a Runaway Train: Errant Economists, Shameful Spenders, and a Plan to Stop them All*, Berkeley: University of California Press.
Czech, B. (2013) *Supply Shock: Economic Growth at the Crossroads and the Steady State Solution*, Canada: New Society Publishers.
Daly, H. (ed.) (1973) *Toward a Steady-state Economy*, San Francisco: W. H. Freeman.
Daly, H. (1991) *Steady State Economics*, Washington: Island Press.
Daly, H. (1996) *Beyond Growth: The Economics of Sustainable Development*, Boston: Beacon Press.
Daly, H. (2008) 'A steady-state economy: a failed growth economy and a steady-state economy are not the same thing; they are the very different alternatives we face', 'thinkpiece' for the Sustainable Development Commission, UK, 24 April 2008, online, available at: http://steadystaterevolution.org/files/pdf/Daly_UK_Paper.pdf (accessed 2 December 2014).
Daly, H. (2012) 'Moving from a failed growth economy to a steady-state economy', in *Towards an Integrated Paradigm in Heterodox Economics*, eds. J. Gerber and R. Stepacher, UK: Palgrave Macmillan.
Daly, H. (2013) 'A further critique of growth economics', *Ecological Economics*, vol. 88, pp. 20–24.
Daly, H. and Cobb, J. (1994) *For the Common Good: Redirecting the Economy toward Community, the Environment, and a Sustainable Future*, Boston: Beacon Press.
Das, S. (2011) *Extreme Money: Masters of the Universe and the Cult of Risk*, New Jersey: FT Press.

Diesendorf, M. (2014) *Sustainable Energy Solutions for Climate Change*, London: Earthscan (Routledge).

Diesendorf, M. and Hamilton, C. (eds) (1997) *Human Ecology, Human Economy*, Sydney; Allen and Unwin.

Dietz, R. and O'Neill, D. (2013) *Enough is Enough: Building a Sustainable Economy is a World of Finite Resources*, San Francisco: Berrett-Koehler Publishers.

Douthwaite, R. (1999) *The Growth Illusion: How Economic Growth has Enriched the Few, Impoverished the Many and Endangered the Planet*, Canada: New Society Publishers.

Economist (2013) Definition of 'economics', online, available at: www.economist.com/economics-a-to-z/e (accessed 18 August 2013).

Ekins, P. (2009) 'Reconciling economic growth and environmental sustainability', paper presented at the Complexity Economics for Sustainability Seminar, Cambridge, 3–4 December 2009.

Elliston, B., MacGill, I. and Diesendorf, M. (2013) 'Least cost 100% renewable electricity scenarios in the Australian National Electricity Market', *Energy Policy*, online, available at: www.ies.unsw.edu.au/sites/all/files/profile_file_attachments/LeastCostElectricityScenariosInPress2013.pdf (accessed 18 August 2013).

Ellul, J. (1975) *The New Demons*, New York: Seabury Press.

EMF (2012) *Towards a Circular Economy: Economic and Business Rationale for an accelerated transition*, Isle of Wight, UK: Ellen Macarthur Foundation, online, available at: www.ellenmacarthurfoundation.org/business/reports/ce2012 (accessed15 July 2014).

Faber, M. (2008) 'How to be an ecological economist', *Ecological Economics*, vol. 66, no. 1, pp. 1–7.

Fletcher, G. (1989) *The Keynesian Revolution and Its Critics: Issues of Theory and Policy for the Monetary Production Economy*, London: Palgrave Macmillan.

Foxon, T., Köhler, J., Michie, J., and Oughton, C. (2012) 'Towards a new complexity economics for sustainability', *Cambridge Journal of Economics*, vol. 37, pp. 187–208.

Georgescu-Roegen, N. (1971) *The Entropy Law and the Economic Process*, Cambridge, MA: Harvard University Press.

Gilder, G. (1981) *Wealth and Poverty*, New York: Bantam.

Godfaurd, J., Clements-Croome, D., and Jeronimidis, G. (2005) 'Sustainable building solutions: a review of lessons from the natural world', *Building and Environment*, vol. 40, no. 3, pp. 319–328.

Heinberg, R. (2011) *The End of Growth: Adapting to Our New Economic Reality*, Canada: New Society Publishers.

Hubbert, K. (1993) 'Exponential growth as a transient phenomenon in human history', in *Valuing the Earth: Economics, Ecology, Ethics*, eds., H. Daly and K. Townsend, Cambridge, MA: MIT Press.

Jackson, T. (2009) *Prosperity Without Growth: Economics for a Finite Planet*, London: Earthscan.

Keynes, J. M. (1936) *The General Theory of Employment, Interest and Money*, London: Palgrave Macmillan.

Klugman, P. (1997) 'Earth in the balance sheet: economists go for green', online, available at: http://web.mit.edu/krugman/www/green.html and http://dieoff.org/page105.htm (accessed 2 February 2012).

Kopnina, H. and Blewitt, J. (2015) *Sustainable Business: Key Issues*, London: Routledge.

Kumar, P. (2010) *The Economics of Ecosystems and Biodiversity: Ecological and Economic Foundations*, London: Earthscan.

Latouche, S. (2010) 'Growing a degrowth movement' in *State of the World 2010: Transforming Cultures from Consumerism to Sustainability*, eds. L. Starke and L. Mastny, Box 22 (p. 181), New York: Worldwatch Institute/Earthscan.

Layard, R. (2005) *Happiness: Lessons from a New Science*, New York: Penguin Press.
Lowe, I. (2005) *A Big Fix: Radical Solutions for Australia's Environmental Crisis*, Melbourne: Black Inc.
Matthews, E., Amann, C, Bringezu, S., Fisher-Kowalski, M., Hutler, W., Kleijn, R., Moriguchi, Y., Ottke, C., Rodenberg, E., Rogich, D., Schandl, H., Schutz, H., Vandervoet, E. and Weisz, H. (2000) *The Weight of Nations: Material Outflows from Industrial Economies*, Washington: World Resources Institute, online, available at: http://pdf.wri.org/weight_of_nations.pdf (accessed 20 July 2014).
Meadows, D., Meadows, D., Randers, J. and Behrens, W. (1972) *The Limits to Growth*, Washington: Universe Books.
Meadows, D., Randers, J and Meadows D. (2004) *The Limits to Growth: The 30-year Update*, Vermont: Chelsea Green.
Meyer, A. (2000) *Contraction and Convergence: The Global Solution to Climate Change*, Schumacher Briefings 5, Devon, UK: Green Books.
Mill, J. S. (1952 [1859]) *On Liberty*, Chicago: Encyclopedia Britannica Great Books.
Mishan, E. (1967) *The Costs of Economic Growth*, New York: F. A. Praeger.
Næss, P. (2011) 'Unsustainable growth, unsustainable capitalism', *Journal of Critical Realism*, vol. 5, no. 2, pp. 197–227.
NEF (2009) 'The Great Transition: how it turned out right', New Economics Foundation, online, available at: www.neweconomics.org/publications/entry/the-great-transition (accessed 18 August 2013).
Nordhaus, W. and Tobin, J. (1973) 'Is growth obsolete?', in *The measurement of Economic and Social Performance*, ed. M. Moss, New York: National Bureau Economic Research.
Oreskes, N. and Conway, M. (2010) *Merchants of Doubt: How a Handful of Scientists Obscured the Truth on Issues from Tobacco Smoke to Global Warming*, New York: Bloomsbury Press.
Palley, T. (2014) 'Making finance serve the real economy', in *State of the World 2014: Governing for Sustainability*, ed. L. Mastny, Washington: Island Press.
Pauli, G. (2010) *The Blue Economy*, Boulder, CO: Paradigm Publishers.
Polanyi, K. (1944) *The Great Transformation*, New York: Rinehart.
Porritt, J. (2005) *Capitalism as if the World Matters*, London: Earthscan.
Prugh, T. (2013) 'Getting to true sustainability', in *State of the World 2013: Is Sustainability Still Possible?*, ed. L. Starke, Washington: Island Press.
Pyarelal, N. (1958) *Mahatma Gandhi: The Last Phase (Volume 10)*, Ahmedabad: Navajivan Publishing House.
Rees, W. (2010) 'What's blocking sustainability? Human nature, cognition and denial, *Sustainability: Science, Practice and Policy*, vol. 6, no. 2 (ejournal), online, available at: http://sspp.proquest.com/archives/vol. 6iss2/1001–012.rees.html (accessed 21 June 2014).
Renner, M. and Prugh, T. (2014) 'Failing governance, unsustainable planet', in *State of the World 2014: Governing for Sustainability*, ed. L. Mastny, Washington: Island Press.
REN21 (2013) *Renewables 2013 Global Status Report*, Renewable Energy Policy Network for the 21st Century, online, available at: www.ren21.net/REN21Activities/GlobalStatusReport.aspx (accessed 3 October 2013).
Robertson, D. (1956 [1954]) 'What does the economist economize?', in *Economic Commentaries*, ed. D. Robertson, London: Staples Press.
Rolston III, H. (2012) *A New Environmental Ethics: The Next Millennium of Life on Earth*, London: Routledge.
Sachs, W. (2013) 'Development and decline', in *State of the World 2013: Is Sustainability Still Possible?*, ed. L. Starke, Washington: Island Press.
Schumacher, E. (1973) *Small is Beautiful*, New York: Harper Torchbooks.

Siegel, C. (2006) *The End of Economic Growth*, Berkeley: The Preservation Institute, online, available at: http://ase.tufts.edu/gdae/CS/EndGrowth.pdf (accessed 2 December 2014).

Simms, A., Johnson, V. and Chowla, P. (2010) *Growth Isn't Possible: Why We Need a New Economic Direction*, London: New Economics Foundation, online, available at: www.neweconomics.org/publications/growth-isnt-possible (accessed 8 March 2012).

Simon, J. (1981) *The Ultimate Resource*, New Jersey: Princeton University Press.

Smith, A. (1759) *The Theory of Moral Sentiments*, London: A. Millar.

Smith, A. (1776) *An Inquiry into the Nature and Causes of the Wealth of Nations*, London: W. Strahan and T. Cadell.

Stahel, W. (2010) *The Performance Economy*, London: Palgrave Macmillan.

Staples, W. and Cafaro, P. (2012) 'For a species' right to exist', in *Life on the Brink: Environmentalists Confront Overpopulation*, eds. P. Cafaro and E. Crist, Georgia, US: University of Georgia Press, pp. 283–300.

Stiglitz, J. (1979) 'A neoclassical analysis of the economics of natural resources', in *Scarcity and Growth Reconsidered*, ed. K. Smith, Baltimore: John Hopkins University Press.

Stiglitz, J. E. (2002) *Globalization and its discontents*, New York: Allen Lane.

Sukhdev, P. (2010) 'Preface' to *The Economics of Ecosystems and Biodiversity: Ecological and Economic Foundations*, ed. P. Kumar, London: Earthscan.

Sukhdev, P. (2013) 'Transforming the corporation into a driver of sustainability', in *State of the World 2013: Is Sustainability Still Possible?*, ed. L. Starke, Washington: Island Press.

Sullivan, A. and Sheffrin, S. (2003) *Economics: Principles in Action*, Upper Saddle River: Pearson Prentice Hall.

Turner, G. (2011) 'Consumption and the environment: impacts from a system perspective', in *Landscapes of Urban Consumption*, ed. P. Newton, Collingwood: CSIRO Publishing.

UNEP (2011a) *Towards a Green Economy: Pathways to Sustainable Development and Poverty Eradication*, United Nations Environment Programme, online, available at: www.unep.org/greeneconomy (accessed 2 December 2014).

UNEP (2011b) *Recycling Rates of Metals: A Status Report*, Nairobi: United Nations Environment Programme.

Victor, P. (2008) *Managing Without Growth: Slower by Design, not Disaster*, Cheltenham, UK: Edward Elgar.

Weizsäcker, E. von, Hargroves, K., Smith, M., Desha, C. and Stasinopoulos, P. (2009) *Factor 5: Transforming the Global Economy through 80% Increase in Resource Productivity*, London: Earthscan.

Welzer, H. (2011) *Mental Infrastructures: How Growth Entered the World and our Souls*, Berlin: Heinrich Böll Foundation.

WGBU (2011) *World in Transition: A Social Contract for Sustainability*, German Advisory Council on Global Change (WBGU).

Whitehead, A. (1929) *Process and Reality*, New York: Harper Brothers.

Wijkman, A. and Rockstrom, J. (2012) *Bankrupting Nature: Denying our Planetary Boundaries*, London: Routledge.

WWF (2011) 'The energy report: 100% renewable energy by 2050', Switzerland: WWF, online, available at: http://wwf.panda.org/what_we_do/footprint/climate_carbon_energy/energy_solutions/renewable_energy/sustainable_energy_report/ (accessed 3 February 2012).

Worster, D. (1988) *The Ends of the Earth: Perspectives on Modern Environmental History*, Cambridge: Cambridge University Press.

Zerubavel, E. (2006) *The Elephant in the Room: Silence and Denial in Everyday Life*, London: Oxford University Press.

Zolotas, X. (1981) *Economic Growth and Declining Social Welfare*, New York: New York University Press.

5
ECOLOGICAL SUSTAINABILITY – ESSENTIAL BUT OVERLOOKED

> *As I step outside*
> *Into the wild,*
> *I embrace*
> *A harmony of lives,*
> *Meshing together,*
> *In a serene*
> *Yet changing balance*
> *Of co-evolved equilibrium,*
> *Where the whole*
> *Is far, far greater*
> *Than the sum of its parts.*
> *Stable yet dynamic –*
> *Such exquisite artistry*
> *Of belonging.*
>
> *(From 'Balance', Haydn Washington 2013a)*

What *is* 'ecological sustainability'? At one level it might seem obvious, to sustain the ecosystems that support human society (and all other species). At its most basic, sustainability must be about keeping an 'ecologically sustainable biosphere' where ecological structure and function continues without major (indeed almost overwhelming) disruption. However, this means that achieving ecological sustainability must also be about taking action to solve the Earth's environmental crisis. It means humanity's ecological footprint should not exceed Earth's productive capacity (GFN 2013). There are many key problems that contribute to our environmental crisis, and they often inter-relate. 'Ecological sustainability' should consider these also if we are to retain a sustainable biosphere. The scope of 'ecological sustainability' thus needs to be quite broad. In terms of urgent challenges, it means solving the climate crisis (Washington and Cook 2011). Another key problem (which scientists don't fully understand the ramifications of) is the ongoing toxification of the planet (Rockstrom *et al.* 2009), with around 100,000 new chemicals commonly in use that didn't exist a century ago.

However, the extinction crisis is overwhelming, as we are now in the midst of the sixth mass extinction we know of in the last 600 million years (Kolbert 2014). The scale of what we are doing, the sheer 'moral evil' (Crist 2012) is almost unimaginable. It was bad enough that in 2003 biodiversity expert E. O. Wilson warned that without action, by the end of the century, *half* of all species on Earth may be extinct (Wilson 2003). It is even more appalling that a 2011 overview (Raven *et al.* 2011) indicated that *two-thirds* of terrestrial species were likely to become extinct by 2100. This will close off the evolutionary potential of more than half the living world. As Soulé and Wilcox (1980) have observed: 'Death is one thing – an end to birth is something else'. We are losing our own family (Crist 2012). Yet how many of the public understand the appalling reality of what we face? How often do the media explain it? Crist (2012, p. 142) notes: 'The ongoing and escalating genocide of non-humans is shrouded in silence, a silence signifying disregard for the vanquished. Silence is how power talks down to the subjugated. Silence is how power disdains to talk about their extinction'.

Do we have a problem?

Apart from this section, I shall not seek to 'prove' the environmental crisis. So many books have summarised 'ecocide', the wiping out of the ecosystems upon which we ourselves depend. It is however worth summarising that during the twentieth century (most figures from Rees 2008):

- Human population increased fourfold to 6.4 billion.
- Industrial pollution went up 40-fold.
- Energy use increased 16-fold and CO_2 emissions 17-fold.
- Fish catches went up 35-fold.
- Water use increased ninefold.
- Mining of ores and minerals grew 27-fold (UNEP 2011).
- One-quarter of coral reefs were destroyed and another 20 per cent degraded (MEA 2005) (75 per cent are now endangered, Postel 2013).
- 35 per cent of mangroves were lost (in just two decades) (MEA 2005).
- At least half of all wetlands were lost to dredging, filling, draining and ditching (Meadows *et al.* 2004, p. 85).
- Extinction is *at least* 1,000-fold above the normal levels in the fossil record (MEA 2005).

Humans use 75 per cent of the Earth's land mass (excluding Greenland and Antarctica) (Erb *et al.* 2009). Some 30 million ha of tropical forest are cleared each year (Daily and Ellison 2002). Half the world's people live in countries where water tables are falling as aquifers are depleted (Brown 2011). Regarding climate change, without action, between 18 and 35 per cent of plant and animal species could be committed to extinction by 2050 (Thomas *et al.* 2004). There are of course the other environmental impacts of climate change, such as: rising sea level; increasing extreme weather; major impacts on water availability and food production; and ocean acidification changing marine ecosystems (Pittock 2009; IPCC 2014). I will not delve deeply into the climate crisis, given that so much has been written, with the IPCC (2014) Fifth Assessment Report updating the extent of the crisis, where human-caused climate change is changing the world rapidly – to the disadvantage of the ecosystems we rely on, and hence our own great disadvantage.

All of the above shows us that we are *bankrupting* Nature (Wijkman and Rockstrom 2012). On a finite world with expanding population and consumption, clearly something has got to give. You cannot keep increasing the stresses on Nature without what Winston Churchill (regarding a different sort of procrastination) called 'a period of consequences'. This is why in 1992, 1,700 of the world's leading scientists published a 'Warning to Humanity' (UCS 1992) that stated:

> Human beings and the natural world are on a collision course. Human activities inflict harsh and often irreversible damage on the environment and on critical resources. If not checked, many of our current practices put at serious risk the future that we wish for human society and the plant and animal kingdoms, and may so alter the living world that it will be unable to sustain life in the manner that we know. Fundamental changes are urgent if we are to avoid the collision our present course will bring about.

This warning was echoed in 2005 by the Millennium Ecosystem Assessment (MEA 2005):

> At the heart of this assessment is a stark warning. Human activity is putting such a strain on the natural functions of the Earth that the ability of the planet's ecosystems to sustain future generations can no longer be taken for granted.

As Wijkman and Rockstrom (2012) conclude, we are consuming the past, present and future of our biosphere. Sustainability requires that our emphasis shift from 'managing resources' to managing *ourselves* that we learn again to live as part of Nature. If we do this, then 'economics at last becomes human ecology' (Wackernagel and Rees 1996). So we do indeed have a problem, which is why we desperately need a 'sustainability' that solves the environmental crisis. A key part of thinking sustainably it to acknowledge our dependence on Nature.

Human dependence on Nature

I have previously written the book 'Human Dependence on Nature' (Washington 2013b) about the ways humans are reliant on Nature. These will be summarised briefly.

74 Ecological sustainability

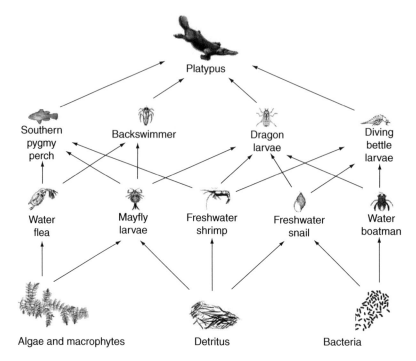

FIGURE 5.1 Simple food web for the Australian platypus (source: adapted from Mount Gambier High School website, online, available at: www.mghs.sa.edu.au/internet/curriculum/science/Year9/livingTogether.htm).

Food webs: Aldo Leopold (1949) characterised the economy of Nature as a 'fountain of energy flowing through a circuit of soils, plants and animals'. Energy is continually being added to ecosystems from the Sun, being trapped by plants (producers) and these plants are eaten by herbivores, which may be eaten by carnivores. This flow of energy in an ecosystem is called a food or 'trophic' chain, and the flow of energy through a web of different species is a *food web*. At each step in the chain, usable energy is lost, so the next step in the food chain has only about 10 per cent of the biomass of the step before it. This is why most food chains have only four or five steps, as the amount of energy remaining after this is too small to support a viable population.

Energy is life: Our planet is bathed in a life-giving stream of energy from the Sun. This supplies the energy for our fields, farm animals, and our own bodies. The fossil fuels that power our civilisation are carbon compounds trapped by photosynthesis in past ecosystems, and fossilised hundreds of millions of years ago. Plants trap 1–2 per cent of the sunlight that falls on them (Hall and Rao 1999). This powers almost all of the Earth's ecosystems, the only exceptions being 'chemoautotrophic' bacteria that derive their energy from converting energy-rich chemicals in rocks (such as sulphur). So the ancient Egyptians had it right, the Sun is the *font of life*.

The Second Law of Thermodynamics tells us that energy goes from a usable form to an unusable form as 'entropy' (disorder) increases. Energy thus passes from a high energy state to a low energy state (heat). It goes in one direction, so unlike nutrients, you cannot 'recycle' energy. The direction can be reversed by human action (e.g. a fridge), but this takes added energy. Thus, ultimately, the amount of life is determined by the fixed amount of sunlight falling on the Earth. Unlike our society's energy consumption, solar energy is not increasing exponentially. We need to discuss some key ecological terms. Gross Primary Productivity (GPP) is the total amount of carbon fixed by plants through photosynthesis. Solar energy is trapped in sugars (and then used to make other compounds). However, plants use some of the sugars to run their own cells (respiration), which uses around half of the energy fixed in sugars (Cain *et al.* 2008). The remainder of the fixed carbon is called Net Primary Productivity (NPP). NPP represents the energy (stored in organic matter) left over for plant growth (and for consumption by herbivores). NPP is thus the basis of food webs that make up the Earth's web of life. More NPP means more opportunities for life, less NPP means less. It is the foundation for the living world, but one of fixed size. We can degrade ecosystems (e.g. by land-clearing) so that we have less foundation, but we cannot easily build 'more' of this foundation than sunlight (and the limits of photosynthesis) will allow.

The energy limits of the Earth's ecosystems cannot be ignored. On a finite Earth we cannot keep increasing the amount of food, fibre and wood we produce. Energy is life, but the amount coming to Earth is fixed. Humanity is now using about 12,000 times as much energy per day as was the case when farming first started: 90 per cent of this is due to industrialisation, 10 per cent to our huge growth in numbers (Boyden 2004). So how much energy is humanity's due? How much of the Earth's productivity should be controlled by just one species? Vitousek *et al.* (1986) estimated that about 40 per cent of NPP in terrestrial ecosystems was being co-opted by humans each year. Rojstaczer *et al.* (2001) argued this could be as high as 55 per cent, while Haberl *et al.* (2007) estimated a figure of 24–29 per cent. Whatever figure one uses, this is a huge percentage of the planet's NPP. How much is enough, how much too much? Of course, if we actually tried for 100 per cent of NPP, then natural ecosystems would completely collapse everywhere, as would civilisation. The fact that 60 per cent of ecosystem services are now being degraded or used unsustainably (MEA 2005) shows our current appropriation of NPP is too high. Humanity's consumption of biomass is almost 100 times higher than the highest level of biomass appropriation by the 96 next highest consuming mammal species (Fowler and Hobbs 2003). Clearly, we are way beyond what could be considered 'equitable' in terms of our fair share. The energy of all ecosystems cannot end up being 'just for us'. Yet to date humanity has gorged itself to bursting point as an NPP glutton, perhaps without fully understanding the consequences.

Keystone species: Each species in a food web may not have equal impact on other species. Some species have more effect on how energy moves through a food web, and even on what species are present. These are known as 'keystone species', important but little known parts of ecosystems. There are three types: 'predators', 'mutualists' and 'ecosystem engineers'. Keystone predators are often found at high levels in the food web, but are not necessarily the top predator. Some are indeed top predators, such as wolves, dingos and jaguars (Nowell and Jackson 1996; Purcell 2010). However, the sea otter and sea star are keystone species not found at the top of the food chain. Keystone mutualists are organisms that participate in mutually beneficial interactions with other organisms, and their loss impacts strongly upon ecosystems. In the Avon wheat belt of Western Australia, at one time of year *Banksia prionotes* is the sole source of nectar for honeyeaters (key pollinators). Therefore the loss of this species would probably cause the honeyeater population to collapse, with profound implications. Other examples of mutualist keystone species are fruit-eaters such as the cassowary, which spread the seeds of many different trees around the rainforest (Walker 1995). It is thus critical we keep keystone species, the trouble is we don't know what many (perhaps most) of them *are*.

Nutrient cycles: Energy may flow through to us from the Sun, but the Earth is finite, with material limits. All life on Earth requires water and minerals to survive. The water that makes up 70 per cent of our bodies; the phosphorus incorporated in our bones and the ATP molecule that powers our cells; the nitrogen in amino acids that form proteins; the potassium we need for cellular reactions and osmotic control. Even the very carbon we eat in food and breathe out as carbon dioxide. If we only used these once, then they would have run out long ago, and life would have faded away. Instead, they are part of the 'great cycles', where each is taken up and used by plants and animals and then returned to the Earth. The hydrologic and sedimentary cycles are intertwined with the distribution of six important elements: hydrogen, carbon, oxygen, nitrogen, phosphorus and sulphur. These make up 95 per cent of all living things (Daily 1997).

Nitrogen is a key nutrient cycle, for it forms the basis of the amino acids that make up proteins. Figure 5.2 shows the nitrogen cycle. Nitrogen gas in the atmosphere is not biologically available, and must be turned into nitrates that plants can absorb. This can happen naturally by the action of lightning, or by nitrogen-fixing bacteria found mostly in plant root nodules (especially legumes). Animals take up nitrogen in proteins by eating plants (or other animals), and animal waste is decomposed back to nitrates. There is thus a great cycle of nitrogen moving from the atmosphere to the land, into living things, then back to ecosystems and back to the atmosphere. Humans now want more for agriculture, so we produce vast amounts of nitrate fertilisers, using fossil fuels. These provide the nutrients to boost production and create the 'ghost acreage' where more food can be produced – as long as you have the fossil energy to make the nitrate fertilisers (Catton 1982).

Producing massive amounts of nitrate fertiliser has consequences. Vitousek *et al.* (1997) concluded that human alterations have approximately doubled the rate of nitrogen input into ecosystems and increased concentrations of the potent greenhouse gas nitrous oxide (N_2O). Over the past four decades, excessive nutrient loading has become one of the most important stresses on ecosystems. Nitrate pollution has already exceeded planetary limits (Rockstrom *et al.* 2009). Phosphorus is another key nutrient cycle, being part of our bones and the vital cellular energy

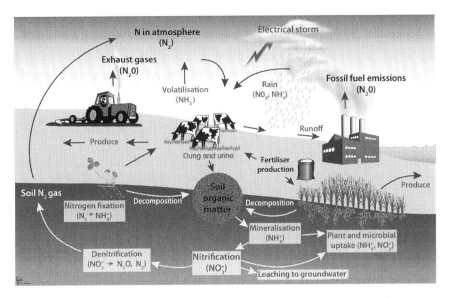

FIGURE 5.2 The global nitrogen cycle (source: Max Oulton, University of Waikato, Hamilton, New Zealand; see Max Oulton's University of Waikato webpage, online, available at: www.waikato.ac.nz/fass/about/staff/maximus).

molecule ATP. It is one of the key limiting elements for plant growth, so humanity has altered the cycle by mining phosphate-rich rocks and guano deposits. This has at least doubled the amount of phosphorus moving through ecosystems. This huge addition of phosphorus puts a major strain on aquatic ecosystems, where it is the key limiting nutrient. Phosphorus pollution can cause massive algae growth and eutrophication, killing fish and sometimes flipping ecosystems into alternative, less diverse but stable states (MEA 2005). Nutrient cycles thus form a key backbone of the life processes that run our world. They deserve our respect, and we disturb them at our peril – yet we continue to do so.

Ecosystem services: Gretchen Daily (1997) states that ecosystem services are conditions and processes through which natural ecosystems 'sustain and fulfil human life'. They maintain biodiversity and the production of ecosystem goods that include seafood, forage, timber, fibre and medicines. They embody the actual life-support functions, such as cleansing and recycling. They also confer important aesthetic, spiritual and cultural benefits. Ecosystem services are commonly defined as the direct and indirect contributions of ecosystems to human well-being (De Groot *et al.* 2010). The Millennium Ecosystem Assessment was written by 1,360 experts (MEA 2005). It split ecosystem services into four parts, being *provisioning services* (products obtained from ecosystems), *regulating services* (benefits obtained from the regulation of ecosystem processes), *cultural services* (non-material benefits) and *supporting services* (those necessary for the production of all other ecosystem services).

The MEA (2005) noted that human use of all ecosystem services is growing rapidly. Overall it concluded that 60 per cent of ecosystem services are being

78 Ecological sustainability

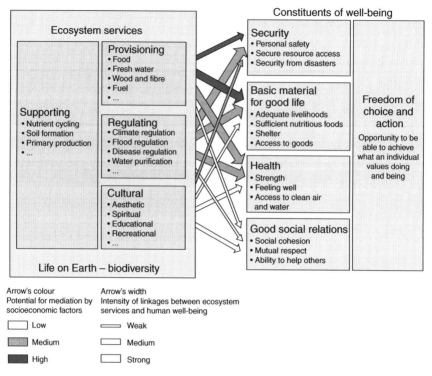

FIGURE 5.3 The relationships between ecosystem health and human well-being (source: Millennium Ecosystem Assessment (MEA 2005), online, available at: www.millenniumassessment.org).

degraded or *used unsustainably*. Many ecosystem services are being degraded primarily to increase food supply.

Ecosystem collapse: Ultimately, human existence depends on maintaining the rich web of life within which we evolved. The rapidity of current environmental change is *unique in human history* (Gowdy *et al.* 2010). Many entire ecosystems are degraded and verging on collapse. This situation has been described as the 'rivet popper' analogy by Ehrlich and Ehrlich (1981). If you keep taking out rivets from an aeroplane, at some point it will fall apart and crash. If you keep removing species, so will ecosystems. Historically, there are no recorded cases where sudden unexpected losses of biodiversity in major ecosystems have been to the benefit of humanity. We know what happens to a house if you take out too many of its foundations, it collapses. The same thing can happen with ecosystems, and the term is entirely 'appropriate' to our situation, so we shouldn't hide this under jargon. Scientists often do use jargon however, and one sees descriptions of 'irreversible non-linear changes' or 'regime shifts'. What they really mean is *collapse*. Such collapses can produce large unexpected changes in ecosystem services. Examples include massive changes in lakes, degradation of rangelands, shifts in fish stocks, breakdown of coral reefs and extinctions due to persistent drought (MEA 2005).

But is ecosystem collapse happening? Well, the MEA (2005) identified that 60 per cent of ecosystem services are being degraded, and found that extinction rates were 1,000 times higher than those in the fossil record. Humanity's ecological footprint doubled in the last 40 years (Sukhdev 2010) and is now at least 1.5 Earths (GFN 2013). This is the amount of land it would take to *sustainably* provide for the demands humanity places upon the Earth. Of course we only have *one* Earth, hence why ecosystems are degrading. We have overshot what the Earth can sustainably provide (Catton 1982). Between 1970 and 2007, the Living Planet Index fell by 28 per cent. This global trend suggests we are degrading natural ecosystems at a rate unprecedented in human history (WWF 2012). According to the IUCN, a quarter of mammal species face extinction (Gilbert 2008). Conservative estimates indicate that 12 per cent of birds are threatened and over 30 per cent of amphibians and 5 per cent of reptiles. Wilson (2003), the 'father' of term biodiversity, says that (without action) by the end of the century, half the planet's species will be extinct.

So what causes an ecosystem to collapse? There are many stresses – loss of habitat, introduced species, nutrient pollution, toxification of ecosystems. The climate crisis on top of all the other environmental problems will accelerate ecosystem decline and collapse. An ecosystem may change gradually until it reaches a threshold and collapses. There can be 'irreversible regime shifts', and it is impossible to know exactly where the dangerous thresholds lie (Elmqvist *et al.* 2010). Once an ecosystem has collapsed, recovery to the original state may take decades or centuries, and may sometimes be impossible. Given the current human impacts on the biosphere, the notions of 'restoration and restorability' need to 'be expressed with many caveats' (Elmqvist *et al.* 2010). In other words, we would be fools to think we can restore all ecosystems that collapse. Some of these non-linear changes can be very large and have substantial impacts on human well-being. The MEA notes that for most ecosystems, science cannot predict the thresholds where change will occur. Thus we cannot say which particular ecosystems will collapse, at what level of stress, or when.

Ecosystem collapse has been caused by: eutrophication; overfishing; species introduction; nutrient impact on coral ecosystems; regional climate change; unsustainable bushmeat trade and loss of keystone species (MEA 2005). Does ecosystem collapse matter to us? There is a real (although unknown) possibility that net biodiversity loss will have catastrophic effects on human welfare (Gowdy *et al.* 2010). Certainly, we do know that humans are totally dependent on the ecosystem services that Nature provides. If enough ecosystems collapse, this must impact big-time on human well-being. So humanity is dependent on Nature, and we have ignored and denied this reality, to our (and Nature's) great cost. Of course, as the old bumper sticker observed, given our obligate dependency – 'Nature bats last'.

Theory and the 'balance of Nature'

'Ecological sustainability' is also affected by how we *think* about Nature and the theories that ecologists and academics develop to explain what they observe.

Problems with theory

There are problems with 'theory', and the ideology that influences it. Modernism and postmodernism are discussed in Chapters 1 and 8 respectively. It is important however to note that science (like Western society and most of academia) is *highly anthropocentric*. It is also true that many scientists don't realise this. 'The Economics of Ecosystems and Biodiversity' (TEEB) project of UNEP is an illustration of how common (and insidious) anthropocentrism can be (Elmqvist *et al.* 2010, p. 44), where a 'key message' states: 'All ecosystems are shaped by people, directly or indirectly and all people, rich or poor, rural or urban, depend on the capacity of ecosystem to generate essential ecosystem services. In this sense people and ecosystems are interdependent social-ecological systems'.

'Shaping' is an anthropocentric term, and implies 'knowing' what you are doing, and being in control. The environmental crisis clearly shows that we don't know what we are doing. It is out of control. Humans don't 'shape' Nature, they *influence* it, and that is a critical difference (see Chapter 8). The final sentence above speaks of 'interdependence', just as the World Conservation Strategy (WCS 1980) did, implying that Nature is also dependent on humanity. However, people rely on ecosystems, they *don't* rely on us. It is hubris to think otherwise. This 'key message' thus actually mis-states the ecological reality of how the world works. TEEB goes on to quote Descola (1996), who states that Nature is 'less and less a product of an autonomous principle of development' and 'its foreseeable demise as a concept will probably close a long chapter of *our own history*' (my emphasis). This is part of what can be called 'Nature scepticism', which suggests Nature is just part of human culture (see Chapter 8). This neatly emphasises the anthropocentrism deeply rooted in academia. This inevitably influences the ideas and theory that academics develop.

The balance of Nature – the case 'for'

One key question we need to consider is the question of 'The balance of Nature'. Does it exist, is it important, and is it in fact what ecological sustainability strives to maintain? To grapple with this, we will have to consider ecological theory. The previous section has touched on energy flow and nutrient cycling, the backbones of ecosystem function. Ecosystems persist over time and perform the ecosystem services that support society (Kumar 2010). The idea of a 'balance of Nature' has been a central theme of thinking about Nature since records were first kept. It has been described as the 'Sacred Balance' (Suzuki and MacConnell 1999). As Chapter 1 showed, ideas of 'harmony', balance and responsibility were (and still are) central to the views of indigenous societies over time. They were (and are) central to ancient and modern Nature-writers. The idea was also central in the history of ecological thought. Frederick Clements described the successional development of ecological communities as comparable to the development of individual organisms (Clements 1916). Later ecologists developed that idea and considered that the ecological community was a 'superorganism' (Hagen 1992).

'Stability' has been much-discussed in ecology, with debate about whether complex (more diverse) food webs (high biodiversity) are more *stable* than simpler food webs. 'Stability' is usually defined as the tendency of the community to remain the same in structure and function over time (Cain *et al.* 2008). The term is now avoided by some ecologists. TEEB (Kumar 2010) talks about high biodiversity leading to 'less variability in functioning' rather than calling it 'stability'. Stability is usually gauged by the size of the changes in organism populations over time. A less stable food web means greater potential for extinction. The question of greater diversity leading to stability has been hotly contested within ecology, with Odum (1953) and Elton (1958) believing this was the case, while May (1973) used food web models to suggest that food webs with high diversity are less stable.

However, rainforests and coral reefs do have high biodiversity and do indeed persist over time. Species diversity and species composition are important in determining the stability in a protozoan food web (Cain *et al.* 2008). Diversity can be expected on average to give rise to ecosystem stability. Hooper *et al.* (2005) conclude that increased biodiversity leads to an increase in 'response diversity' (range of traits of species in the same functional groups in response to environmental drivers or stresses), resulting in less variability in function over time. Also, more biodiverse ecosystems have: greater productivity; greater drought tolerance; better water management; better nutrient cycling; greater community respiration; greater biotic resistance to pests; and greater resilience (Cain *et al.* 2008; Elmqvist *et al.* 2010). More diverse ecosystems can absorb up to 30 per cent more CO_2 than monocultures (Daily 1997). More diverse ecosystems thus produce greater ecosystem services. Arguably, more diverse systems are thus better at maintaining the 'balance of Nature'.

Rather than 'stability', the focus has shifted in the last decade to *resilience*. The most common definition is that it represents the capacity of a system to cope with disturbances without shifting into a qualitatively different state. A resilient system has the capacity to withstand shocks and surprises, and if damaged to rebuild itself (Elmqvist *et al.* 2010). Hence 'resilience' is the capacity of a system to both deal with change and continue to develop. A resilient ecosystem is thus also a stable one, one that could be said to 'be in balance', even though that term is rarely used for the reasons described below. In summary, many scholars and ecologists have seen, and still see, a 'balance in Nature', and that humans must act in a way that does not unbalance ecosystems. They believe this is the key aim of ecological sustainability.

The balance of nature – the case 'against'

It has become common in some circles to argue that the 'balance of nature' doesn't exist. Pickett *et al.* (1992) state:

> the classical paradigm in ecology, with its emphasis on the stable state, its suggestion of natural systems as closed and self-regulating, and its resonance with the non-scientific idea of the balance of nature, can no longer serve as an

adequate foundation for conservation. The new paradigm, with its recognition of episodic events, openness of ecological systems and multiplicity of locus and kind or regulation, is in fact a more realistic basis.

The debate is thus about 'equilbrium' or 'dis-equilibrium'. Nega (1998, p. 4) believes 'stability' emphasises constancy and predictability, while a 'resilience' perspective would 'presume the unexpected' in the face of considerable ignorance about the consequences of our actions. Nega says a resilience view, as described by Holling (1973) (the 'father' of 'Adaptive Management' theory), asks how the complexity of ecosystems could be described to better formulate policies that enhance human/ecological sustainability. Gunderson and Holling (2002, p. 34) in their book 'Panarchy' enlarge on the 'adaptive cycle' theory. 'Panarchy' literally means a form of governance that encompasses all others, however they meant by the term that it was the structure in which systems (including those of Nature) are interlinked in continual adaptive cycles of growth, accumulation, restructuring and renewal. They show a 'figure of 8' of the 'adaptive cycle', showing an *exploitation phase*; a *conservation phase*; a *release phase* or 'creative destruction'; and a *reorganisation phase*. There is no explanation as to why it should be a figure of 8 and not a simple circle. Nor is there any explanation as to why this is better than a simple climax/disturbance/recovery cycle. Every field ecologist knows that disturbance (disequilibrium) occurs in ecosystems (fire, storm, etc.), and that ecosystems recover from this, so there can be 'pulses' within natural ecosystems. However, Gunderson and Holling (2002, p. 150) go further to argue:

> There is no nature out there, there is no baseline, current states of nature are seen as extremely path dependent. The environment is not constant and environmental change is episodic. Ecosystems do not have a single equilibrium, but multiple and movement between these is a natural part of maintaining structure and diversity.

They conclude (p. 396) that the adaptive cycle is 'the fundamental unit of dynamic change from cells to ecosystems to cultures' and state that: 'sustainability requires both change and persistence, and is maintained by relationships among a nested set of adaptive cycles arranged in a dynamic hierarchy in space and time – the panarchy'.

'Panarchy' is thus situated as the guiding principle of life, society and seemingly the Universe. However, Whitehead (1929) warned of the 'Fallacy of Misplaced Concreteness', where one abstracts too far from reality. One can only wonder does this apply here? There is a risk of losing real world relevance if one abstracts too far from reality. At a theoretical level 'conservation ecology has been strongly influenced by Holling's Figure 8 Adaptive Cycle Model and cross scale panarchies' (Manuel-Navarrete *et al.* 2008). It is thus an influential theory, and the book 'Panarchy' is referred to as a virtual 'bible' of adaptive management. So the view of the authors on the environmental crisis is worth exploring (Gunderson and Holling

2002, p. 14): 'if human exploitation leads to resource collapse, why haven't all ecosystems collapsed and why are humans still here?'. Resilience is one answer, the other lies in human behaviour and creativity where people can learn and innovate (ibid, p.15).

This supposed 'paradox' seems odd when we are entering the sixth great extinction event (Kolbert 2014), with extinction rates at least 1,000 times those in the fossil record, while 60 per cent of ecosystem services are being degraded (MEA 2005), and our ecological footprint is 1.5 Earths (GFN 2013). There is in fact no 'paradox', many ecosystems *have* collapsed and many others are on the brink of doing so (Kumar 2010). They further claim that 'collapses and the subsequent need to innovate, create, reorganise and rebuild are likely the inevitable consequence of human interactions with nature' (Gunderson and Holling 2002, p. 192). This makes it sound as if ecosystem collapse is *natural* and normal (and that we humans just can't help doing it!). However, it confuses 'natural change' and cycles in ecosystems with the hugely larger (and often previously unknown) stresses that humans are putting on Nature today (e.g. rapid climate change and toxification). In so doing they actually *trivialise* the environmental crisis. They give the impression that 'release' or 'collapse' is natural, and in so doing provide justification for exploiters to argue there is no environmental crisis, and that we should not worry about increasing ecosystem collapse. Later they ask 'is it desirable to have a goal of preserving and protecting systems in a pristine, static state?' (ibid., p. 31). Conservation ecologists and conservationists would reply that protecting wild Nature (a dynamic stability, not a static one) *is* indeed desirable for very good reasons of biogeography and systems ecology (Miller *et al.* 2014). Environmental ethicists would argue it is desirable as part of an 'Earth ethics' (Rolston 2012). However, Gunderson and Holling (2002, p. 99) don't think so:

> Predictions of looming economic crises and collapse caused by resource scarcity are an important part of the sustainability debate, but economist Solow (1973) provides a 'withering critique' of such doomsday scenarios, that they ignore the forward looking behaviour of people.

This argument is repeated four times in 'Panarchy'. The 'doomsday scenario' referred to was the renowned 'Limits to Growth' study (Meadows *et al.* 1972). Gunderson and Holling come down on the side of the neoclassical economist Solow (1974), who we found in Chapter 3 was a champion of 'weak sustainability', the ecologically absurd position that we could swap natural capital for money. However, just because one neoclassical economist was in denial of the environmental crisis, and felt threatened by the 'Limits to Growth' study, does not mean it was false. Turner (2008) of CSIRO has shown that in fact the 'Limits to Growth' standard model has been remarkably accurate over 30 years. It thus is of concern that the key proponents of 'adaptive management theory' seem to be in denial of the Earth's limits, and the environmental crisis caused by us exceeding planetary boundaries (e.g. Rockstrom *et al.* 2009). However, the adaptive management theory epitomised by Gunderson and Holling (2002) is now deeply embedded in environmental circles. It is also a basis for 'resilience thinking', the new buzz word of the last few years. In fact some state that 'debate is raging' within the environment community about whether resilience should replace or augment sustainability as the dominant paradigm (Mazur 2013, p. 361). Given the discussion above, I find this unsettling. Part of that unease comes from four decades of observing ecosystems in wild places, and what they have taught me. Given that all people have a worldview and values, the arguments and theory of Gunderson and Holling seem to be strongly influenced by an apparent neoliberal worldview. My concern is that this theory is influential and often portrayed as a 'given truth', when in fact it seems to have ideological roots. 'Sustainability' needs to be far more than the latest fashion in theory.

The idea that collapse and environmental degradation are just 'natural change' has worried some writers to the extent that Kingsnorth (2013) has christened it 'neoenvironmentalism'. Peter Kareiva (chief scientist of the UK Nature Conservancy) proposes such a view, arguing: 'Humans degrade and destroy and crucify the natural environment, and 80 percent of the time it recovers pretty well' (BI 2012). He argues that trying to protect large functioning ecosystems from human development is 'mostly futile', as humans like development, and 'you can't stop them from having it'. He concludes that 'Nature is tough and will adapt to this'. Similarly, journalist Emma Marris (2011) states:

> For decades people have unquestioningly accepted the idea that our goal is to preserve nature in its pristine, pre-human state. But many scientists have come to see this as an outdated dream that thwarts bold new plans to save the environment and prevents us from having a fuller relationship with nature.

This approach by Kareiva, Marris and others has been called the 'Gardener' or 'new conservation' ethic. It poses a major threat to more than a century of conservation gains (Foreman 2012). Miller *et al.* (2014) note that it comes mostly from those with a neoliberal ideology, and that they usually justify further development on social justice and humanitarian grounds. The 'Gardener' ethic is clearly anthropocentric and utilitarian. Miller *et al.* (2014) note that Gardeners propose a number of wrong (and unethical) premises:

- Nature is just a warehouse for humans.
- We can construct ecosystems out of exotic species.
- We don't have to live within the Earth's limits.
- Nature is more resilient than we thought, so we can continue development.
- Human influence has left no pristine Nature anyway, so this justifies degrading *any* natural area.
- 'Nature' is just a human social construct.
- We can manage Nature for human use and still protect biodiversity *without* keeping large natural areas (wilderness).

Miller et al. explain that the 'new conservation' of the Gardeners is scientifically mistaken and ignores or denies conservation biology. It seeks to use resilience to 'justify massive ecological tinkering'. Thus 'Gardening' is a misrepresentation and twisting of ecological science. It ignores that native biodiversity is the *key factor* in ecosystem health (Tilman 2012). It is part of an ongoing movement to academically justify further destruction of wilderness, similar to the attempts by the 'Wise Use' movement (Washington 2006). Miller et al. (2014) note that increasingly the boards of environmental non-government organisations are dominated by corporate interests whose values are antithetical to the protection of biodiversity 'for its own sake'. Given that Gardeners and neo-environmentalists seem to share a neoliberal ideology, one must ask: 'Are their worldview and beliefs *influencing their theory*?'

Ideology determining theory

Historian Donald Worster (1994) has shown that the history of ecology has been swayed at times by prevailing paradigms (or ideologies). He made a detailed study of equilibria and disequilibria theories within ecology, and pointed out that such theories often tie in with the worldviews of their promoters. Using a principle of 'historicism', he argues we can 'approach recent ecological models that dramatise disturbance with a sense of scepticism and independence' (Worster 1994). He wonders if they are the 'mere reflection of global capitalism and its ideology'. In regard to Nature's dynamism, he concludes:

> Nature, ecologists began to argue, is wild and unpredictable. Nature is in deep, important ways quite disorderly. Nature is a seething, teeming spectacle of diversity. Nature, for all its strange and disturbing ways, its continuing capacity to elude our understanding, still needs our love, our respect, and our help.

Environmental ethicist Holmes Rolston (2012, p. 162) concludes that equilibrium theory and non-equilibrium theory represent 'two ends of a spectrum', with real ecosystems somewhere in between. Whether one sees equilibrium or non-equilibrium can depend on the level and scale of analysis. O'Neill (1986) notes that those who see stability, and those who see change, are looking at two sides of one coin: 'in fact, both impressions are correct, depending on the purpose and time-space scale of our observations'. Pahl-Wostl (1995) concludes 'the dynamic nature of ecosystems is chaos and order entwined'. So there is change and dynamism in ecosystems. Yet there is *also* stability. Just as a population can remain stable, even though it is always changing.

In conclusion, as a field ecologist who has spent decades walking through and studying wild places, I have no problem speaking of Nature's 'dynamic balance'. Of course I see 'change', I also see an ongoing harmony and balance. I feel this every time I step into the wild. We have to be careful about what worldview or ideology influences our view of Nature's 'balance' (see Chapter 8). As Worster (1994) notes, those with a modernist worldview will see Nature as having no balance, as always changing, and thus development is seen as fine, as it is depicted as 'just one more change'. Those with an ecocentric worldview will see the stability, harmony and balance in Nature. However, suggesting that the natural changes within ecosystems (which species co-evolved with) means that humanity can alter the world in ways *far beyond* natural limits is clearly false. Similarly, arguing that we don't need to protect large natural areas (= wilderness), that 'nature is tough and will adapt' is delusional and ignores the current state of knowledge in conservation biology and environmental science. Keeping our remaining wilderness is essential to stop a further cascade of extinction (Mackey *et al.* 1998). Proponents of the Gardener view seek to justify more development of natural areas as 'scientific' when in fact it is *ideological*, an ideology of endless growth and 'mastery of Nature', an ideology that has directly led to the environmental crisis (see Chapter 8). This is to my mind a serious and major twisting of ecological theory.

Why have I spent so much time discussing ecological theory? First, because I am an ecologist, and second because it influences how scientists and academics see and portray the natural world, and what they teach to students. The environmental crisis is critical right now. Any academic teaching of theory that trivialises the current crisis (and supports its ongoing causes) needs to be examined closely. Nature *does* have a 'dynamic balance', and we really have pushed it way beyond its limits, causing collapse. Surely theory that seeks to play down human impact (and which aids denial of the environmental crisis) needs to be modified, or perhaps abandoned?

What *should* ecological sustainability mean?

So what *should* ecological sustainability mean? What is it we are sustaining? Originally this was 'development', and for some that meant sustaining opportunity, sustaining growth, sustaining profits, sustaining capital (artificial or natural), sustaining quality of life, sustaining freedom to choose an abundant life. Historically, ecologists then appeared on the scene, warning us that nobody yet had described sustainability properly in ecological terms. Sadly, this situation largely continues, due to the ignorance and denial of ecological reality. The ultimate goal of any meaningful sustainability must be a 'sustainable biosphere' (Rolston 2012, p. 218). It is clear that part of the answer is not just scientific but is also ethical, and arises from our worldview (see Chapter 8). The scientific part is that we must maintain an ecologically sustainable biosphere into the future. We owe this to future generations of humans, and ethically I believe we owe it to our brother and sister species we share this world with. When discussing the three strands of sustainability, there can be little doubt that society has ignored ecological sustainability more than the other two. Scholars will leap to point out that we have also failed regarding economic and social sustainability. This is true, but at least we have been 'talking' about them. Since Adam Smith, scholars have talked about what economics is, and how to keep it sustainable (however imperfectly). Similarly, we have talked about our society and the best way to reach 'utopia' since the time of Plato. The same is not true for ecological sustainability.

Arguably, most indigenous cultures understood (and some still understand) the need for ecological sustainability, and in Australia this was codified by Aboriginal Elders as 'Law' (Neidjie et al. 1985) and totemism. However, the West has ignored this wisdom for the last two centuries, and globalisation has exported this modernist 'endless growth' view to most other countries. Rachel Carson brought ecological sustainability back to social consciousness in 1962 with 'Silent Spring'. Environmental scientists and environmentalists have been fighting a silence about (and denial of) the environmental crisis ever since. So the first and foremost part of reaching ecological sustainability is to acknowledge that *we must talk about it*, and acknowledge we are dependent on Nature. We must acknowledge that a sustainable biosphere is the absolute non-negotiable *necessity* that underlies sustainability in general (Rolston 2012). Ecological sustainability (at a minimum) should include:

- Increasing our understanding of environmental science and using it to guide our actions.
- Understanding that we depend on Nature to survive, there is no 'planet B' (Brown 2011).
- Controlling the major 'drivers' of environmental degradation: overpopulation and overconsumption (see Chapter 7).
- Confronting and solving the climate crisis by decarbonising our society within decades. This means moving immediately to renewable energy and energy efficiency, as discussed in Chapter 10.
- Reducing the amount of NPP that humanity co-opts and controls. This means no further clearing of natural ecosystems; it means extensive ecological restoration; it means greater connectivity between natural areas; it means curtailing huge developments and controlling urban sprawl. Essentially, it also means controlling population growth and overconsumption (Chapter 7).
- Taking action to return below the safe threshold of the three planetary boundaries we have already exceeded (biodiversity loss, climate change, nitrogen pollution) (Rockstrom et al. 2009).
- Mandating protection of ecosystem services and building this into all economic valuation and planning processes. The 'Constant Natural Capital Rule' of strong sustainability must be applied universally. Actually, we need to go further and adopt the 'strongest sustainability' described in Chapter 3.
- Protecting keystone species, or species likely to be keystone (such as top predators). We cannot wait for research to prove these, for by then they may be extinct. The application of the 'Precautionary Principle' is thus critical for ecological sustainability.
- Stopping further toxification of the planet. This means both limiting resource throughput (see Chapters 4 and 7), and strictly controlling the 100,000 chemicals in use (Rockstrom et al. 2009) and the creation of the 1,000 or so new synthetic chemicals added each year (ATDSR 2010).

- Retaining large natural areas (wilderness), national parks and other reserves, as these are the keystone foundations of conservation, and also the most cost-effective way to keep biodiversity into the future (Mackey et al. 1998). We should aim to keep half of our landscapes set aside for biodiversity protection and evolutionary viability (Schmiegelow et al. 2006).
- Promoting 'Wild Law', Earth jurisprudence and the 'Rights of Nature' and building these into legislation at all levels (Cullinan 2003, 2014). We need a 'Planetary Declaration on the Rights of Nature' (Higgins 2009).
- Managing our 'production landscapes' (farms, producing forests) so as to be ecologically sustainable. This means bringing nutrient cycles into balance through reducing resource throughput, moving to organic farming, and recycling nutrients (rather than mining them). It means a return to organic farming that minimises artificial fertiliser and biocide use and uses Integrated Pest Management (EPA 2014). If done properly this can produce as much food as conventional agriculture (Wijkman and Rockstrom 2012, p. 57). It means re-designing our agriculture to be truly ecologically sustainable (Hill 2014).
- Stopping the world's huge soil erosion, which drains the life-blood of agriculture. Estimates of the net rate of soil losses range from 24 billion tonnes (Pimentel and Pimentel 2008) to 75 billion tonnes (Brown 2011) a year. Some two billion ha or 17 per cent of Earth's vegetated surface have undergone human induced soil degradation since 1945 (Daily 1997, p. 126). Topsoil is being lost 16–300 times *faster* than it can be replaced (Meadows et al. 2004, p. 57), and every year soil erosion causes ten million ha of world cropland to be abandoned (Pimentel and Pimentel 2008).
- Maintaining and improving air and water quality, and allowing significant environmental flows in rivers. This means also not 'mining' groundwater at an unsustainable rate, where in places it is dropping by 5 metres a year (Pimentel and Pimentel 2008, p. 290).
- Not using non-renewable resources if we can switch to renewable ones, and only using these in an ecologically sustainable way. This means not overcutting our forests or overharvesting our fisheries (as we currently do).
- Reducing our overconsumption and moving to Factor 5 (von Weizsäcker et al. 2009), where we use only 20 per cent of the energy and materials we currently use in the West. It means banning 'planned obsolescence' and making 'cradle to cradle' products mandatory in regard to product eco-design (Braungart and McDonough 2008). Container deposit legislation should be mandatory. Plastic use and pollution must be strictly controlled (see Chapter 7). This will require strong regulation.

So ecological sustainability requires many things. Perhaps the most difficult is moving it to 'centre stage'. Humanity can stumble on without 'utopia', and has so far stumbled on with a flawed economic system. However, we cannot continue without an *ecologically sustainable biosphere*. We can no longer take Nature for granted. Ignore it and it will indeed go away. Indeed it is. If sustainability is a three-legged stool, then the leg that is most at risk of breaking catastrophically is ecological sustainability. Yet as a society we still can't quite see this. Or worse, we deny it is even needed (such as the 'Gardeners'). As Crist and Cafaro (2012) conclude:

> We find joy in the abundant beauty and variety that Earth provides. We find safety in a relatively predictable climate and reliable food sources. We find inspiration in the grandeur of this extraordinary planet.... We thus are called, even at this late hour, to find the clarity and courage to shift into a new relationship with Earth, before we diminish irrevocably the greater-than-human world, our own lives, and the lives of future generations.

Ecological sustainability has to be a 'key focus' of sustainability. It is not negotiable, it must be mandatory if we are to reach a sustainable future and a sustainable society. Our ancestors understood this wisdom, but the modern world has lost it, buried it under a modernist mindset of endless growth. It is essential we now recover that old 'Wisdom of the Elders' (Knudtson and Suzuki 1992). Ecological sustainability must be about healing and restoring this wondrous jewel of a planet. The world has ecological limits *we must live within*. This is an essential wisdom both pragmatic and ethical.

References

ATDSR (2010) 'Chemicals, cancer and you' booklet, Agency for Toxic Substances and Disease Registry, online, available at: www.atsdr.cdc.gov/emes/public/docs/Chemicals,%20Cancer,%20and%20You%20FS.pdf (accessed 15 June 2014).
BI (2012) 'So you want to be a conservationist?', Breakthrough Institute, online, available at: http://thebreakthrough.org/archive/peter_kareivas_breakthrough (accessed 18 March 2013).
Boyden, S. (2004) *The Biology of Civilisation: Understanding Human Culture as a Force in Nature*, Sydney: UNSW Press.
Braungart, M. and McDonough, W. (2008) *Cradle to Cradle: Remaking the Way we Make Things*, London: Vintage Books.
Brown, L. (2011) *World on the Edge: How to Prevent Environmental and Economic Collapse*, New York: W. W. Norton and Co.
Cain, M., Bowman, W. and Hacker, S. (2008) *Ecology*, Massachusetts: Sinauer Associates.
Carson, R. (1962) *Silent Spring*, Boston: Houghton Mifflin.
Catton, W. (1982) *Overshoot: The Ecological Basis of Revolutionary Change*, Chicago: University of Illinois Press.
Clements, F. (1916) *Plant Succession: An Analysis of the Development of Vegetation*, Washington, DC: Carnegie Institution of Washington.
Crist, E. (2012) 'Abundant Earth and the population question', in *Life on the Brink: Environmentalists Confront Overpopulation*, eds. P. Cafaro and E. Crist, Georgia: University of Georgia Press, pp. 141–151.
Crist, E. and Cafaro, P. (2012) 'Human population growth as if the rest of life mattered', in *Life on the Brink: Environmentalists Confront Overpopulation*, eds. P. Cafaro and E. Crist, Georgia: University of Georgia Press, pp. 3–15.

Cullinan, C. (2003) *Wild Law: A Manifesto for Earth Justice*, Totnes, Devon: Green Books.
Cullinan, C. (2014) 'Governing people as members of the Earth community', in *State of the World 2014: Governing for Sustainability*, ed. L. Mastny, Washington: Island Press.
Daily, G. (1997) *Nature's Services: Societal Dependence on Natural Ecosystems*, Washington: Island Press.
Daily, G. and Ellison, K. (2002) *The New Economy of Nature: The Quest to make Conservation Profitable*, Washington: Island Press.
De Groot, R., Fisher, B. and Christie, M. (2010) 'Integrating the ecological and economic dimensions in biodiversity and ecosystem service valuation', in *The Economics of Ecosystems and Biodiversity: Ecological and Economic Foundations*, ed. P. Kumar, London: Earthscan.
Descola, P. (1996) 'Constructing natures: Symbolic ecology and social practice', in *Nature and Society: Anthropological Perspectives*, eds. P. Descola, and G. Palsson, New York: Routledge.
Ehrlich, P. and Ehrlich, A. (1981) *Extinction: The Causes and Consequences of the Disappearance of Species*, New York: Random House.
Elmqvist T., Maltby, E., Barker, T., Mortimer, M. and Perrings, C. (2010) 'Biodiversity, ecosystems and ecosystem services' in *The Economics of Ecosystems and Biodiversity: Ecological and Economic Foundations*, ed. P. Kumar, London: Earthscan.
Elton, C. S. (1958) *Ecology of Invasions by Animals and Plants*, London: Chapman & Hall.
EPA (2014) 'Integrated Pest Management (IPM) principles', online, available at: www.epa.gov/opp.00001/factsheets/ipm.htm (accessed 15 June 2014).
Erb, K., Haberl, H., Krausmann, F., Lauk, C., Plutzar, C., Steinberger, J., Muller, C., Bondeau, A., Waha, K. and Pollack, G. (2009) 'Eating the planet: feeding and fueling the world sustainably, fairly and humanely – a scoping study', Social ecology working paper no. 116, Vienna: Institute of Social Ecology and Potsdam Institute for Climate Impact Research.
Foreman, D. (2012) *Take Back Conservation*, Colorado: Raven's Eye Press.
Fowler, C. and Hobbs, L. (2003) 'Is humanity sustainable?', *Proceedings of the Royal Society, Series B: Biological Sciences*, vol. 270, pp. 2579–2583.
GFN (2013) 'World footprint, do we fit on the planet?', Global Footprint Network, online, available at: www.footprintnetwork.org/en/index.php/GFN/page/world_footprint/ (accessed 31 August 2013).
Gilbert, N. (2008) 'A quarter of mammals face extinction', *Nature*, vol. 455, p. 717.
Gowdy, J., Howarth, R. and Tisdell, C. (2010) 'Discounting, ethics and options for maintaining biodiversity and ecosystem integrity', in *The Economics of Ecosystems and Biodiversity: Ecological and Economic Foundations*, ed. P. Kumar, London: Earthscan.
Gunderson, L. and Holling, C. (2002) *Panarchy: Understanding Transformations in Human and Natural Systems*, Washington: Island Press.
Haberl, H., Erb, K., Krausmann, F., Gaube, V., Bondeau, A., Plutzar, C., Gingrich, S., Lucht, W. and Fischer-Kowalski, M. (2007) 'Quantifying and mapping the human appropriation of net primary production in earth's terrestrial ecosystems', *Proceedings of the National Academy of Sciences of the USA*, vol. 104, pp. 12942–12947.
Hagen, J. (1992) *An Entangled Bank: The Origins of Ecosystem Ecology*, New Brunswick: Rutgers University Press.
Hall, D. O. and Rao, K. (1999) *Photosynthesis* (sixth edition), Cambridge, UK: Cambridge University Press.
Higgins, P. (2009) 'Trees have rights too: ecological justice for all' by lawyer Polly Higgins, online, available at: http://treeshaverightstoo.com/ (accessed 14 November 2011).
Hill, S. B. (2014) 'Considerations for enabling the ecological redesign of organic and conventional agriculture: a social ecology and psychosocial perspective', in *Organic*

Farming, Prototype for Sustainable Agricultures, eds. S. Bellon and S. Penvern, Dortrecht: Springer.
Holling, C. (1973) 'Resilience and stability of ecological systems', *Annual Review of Ecology and Systematics*, vol. 4, pp. 1–23.
Hooper, D., Chapin, F., Ewel, J., Hector, A., Inchausti, Lavore, S., Lawton, J., Lodge, D., Loreau, M., Naeem, S., Schmid, Setala, H., Symstad, A., Vandermeer, J. and Wardle, D. (2005) 'Effects of biodiversity on ecosystem functioning: a consensus of current knowledge', *Ecological Monographs*, vol. 75, no. 1, pp. 3–35.
IPCC (2014) 'Summary for Policymakers, Fifth Assessment Report, International Panel on Climate Change', online, available at: http://ipcc-wg2.gov/AR5/images/uploads/WG2AR5_SPM_FINAL.pdf (accessed 16 June 2014).
Kingsnorth, P. (2013) 'Dark ecology: searching for truth in a post-green world', *Orion Magazine*, January/February 2013, online, available at: www.orionmagazine.org/index.php/articles/article/7277 (accessed 21 July 2014).
Knudtson, P. and Suzuki, D. (1992) *Wisdom of the Elders*, Sydney: Allen and Unwin.
Kolbert, E. (2014) *The Sixth Extinction: An Unnatural History*, New York: Holt and Company.
Kumar, P. (2010) *The Economics of Ecosystems and Biodiversity: Ecological and Economic Foundations*, London: Earthscan.
Leopold, A. (1970 [1949]) *A Sand Country Almanac, with Essays on Conservation from Round River*, New York: Random House.
Mackey, B., Lesslie, R., Lindenmayer, D., Nix, H. and Incoll, R. (1998) *The Role of Wilderness in Nature Conservation (Report to Environment Australia)*, Canberra: Centre for Research and Environmental Studies, Australian National University, online, available at: http://secure.environment.gov.au/heritage/publications/anlr/pubs/rolewild.pdf (accessed 21 July 2014).
Manuel-Navarrete, D., Kay, J. and Dolderman, D. (2008) 'Evolution of the ecological integrity debate', in *Sustaining Life on Earth: Environmental and Human Health through Global Governance*, ed. C. Soskolne, New York: Lexington Books.
Marris, E. (2011) *Rambunctious Garden: Saving Nature in a Post-Wild World*, USA: Bloomsbury Publishing.
May, R. (1973) *Stability and Complexity in Model Ecosystems*, Princeton, NJ: Princeton University Press.
Mazur, L. (2013) 'Cultivating resilience in a dangerous world', in *State of the World 2013: Is Sustainability Still Possible?*, ed. L. Starke, Washington: Island Press.
MEA (2005) *Living Beyond Our Means: Natural Assets and Human Wellbeing, Statement from the Board, Millennium Ecosystem Assessment*, United Nations Environment Programme (UNE), online, available at: www.millenniumassessment.org.
Meadows, D., Meadows, D., Randers, J. and Behrens, W. (1972) *The Limits to Growth*, Washington: Universe Books.
Meadows, D., Randers, J and Meadows D. (2004) *The Limits to Growth: The 30-year Update*, Vermont: Chelsea Green.
Miller, B., Soulé, M. and Terborgh, J. (2013) 'The "New Conservation's" surrender to development', Rewilding Institute, online, available at: http://rewilding.org/rewildit/images/The-%E2%80%9CNew-Conservation%E2%80%99s%E2%80%9D-Surrender-to-Development1.pdf, see also: http://onlinelibrary.wiley.com/doi/10.1111/acv.12127/pdf (accessed 23 May 2013).
Nega, T. (1998) 'An annotated bibliography on adaptive environmental assessment and management, 1973–1996', A report of the Northwest Collaboratory for Sustainability, The Institute for Social, Economic and Ecological Sustainability, University of Minnesota.

Neidjie, B., Davis, S. and Fox, A. (1985) *Kakadu Man: Bill Neidjie*, Queanbeyan, NSW: Mybrood.

Nowell, K. and Jackson, P. (1996) *Wild Cats, Status Survey and Conservation Action Plan*, Gland, Switzerland: IUCN/SSC Cat Specialist Group, IUCN.

O'Neill, R., DeAngelis, J., Waide, J. and Allen, T. (1986) *A Hierarchical Concept of Ecosystems*, Princeton: Princeton University Press.

Odum, E. P. (1953) *Fundamentals of Ecology*, Philadelphia: Saunders.

Pahl-Wostl, C. (1995) *The Dynamic Nature of Ecosystems: Chaos and Order Entwined*, New York: Wiley.

Pickett, S., Parker, T. and Fiedler, P. (1992) 'The new paradigm in ecology: implications for conservation biology above the species level', in *Conservation Biology*, eds. P. Fiedler and S. Kain, New York: Chapman and Hall.

Pimentel, D. and Pimentel, M. (2008) 'The future: world population and food security', in *Sustaining Life on Earth: Environmental and Human Health through Global Governance*, ed. C. Soskolne, New York: Lexington Books.

Pittock, A. B. (2009) *Climate Change: the Science, Impacts and Solutions*, Australia: CSIRO Publishing/Earthscan.

Postel, S. (2013) 'Sustaining freshwater and its dependents', in *State of the World 2013: Is Sustainability Still Possible?*, ed. L. Starke, Washington: Island Press.

Purcell, B. (2010) *Dingo*, Canberra: CSIRO Publishing.

Raven, P., Chase, J. and Pires, J. (2011) 'Introduction to special issue on biodiversity', *American Journal of Botany*, vol. 98, pp. 333–335.

Rees, W. (2008) 'Toward sustainability with justice: Are human nature and history on side?', in *Sustaining Life on Earth: Environmental and Human Health through Global Governance*, ed. C. Soskolne, New York: Lexington Books.

Rockstrom, J., Steffen, W., Noone, K., Persoon, A., Chapin, F., Lambin, E., Lenton, T., Scheffer, M., Folke, C., Schellnhuber, H., Nykvist, B., de Wit, C., Hughes, T., van der Leeuw, S., Rodhe, H., Sorlin, S., Snyder, P., Costanza, R., Svedin, U., Falkenmark, M., Karlberg, L., Corell, R., Fabry, V., Hansen, J., Walker, B., Liverman, D., Richardson, K., Crutzen, P. and Foley, J. (2009) 'Planetary boundaries: Exploring the safe operating space for humanity', *Ecology and Society*, vol. 14, no. 2, p. 32, online, available at: www.ecologyandsociety.org/vol.14/iss2/art32/ (accessed 21 July 2014).

Rojstaczer S., Sterling S. and Moore, N. (2001) 'Human appropriation of photosynthesis products', *Science*, vol. 294, no. 5551, pp. 2549–2552.

Rolston III, H. (2012) *A New Environmental Ethics: The Next Millennium of Life on Earth*, London: Routledge.

Schmiegelow, F., Cummings, S., Harrison, S., Leroux, S., Lisgo, K., Noss, R. and Olsen, B. (2006) 'Conservation beyond crisis management: a conservation-matrix model', Beacons Discussion Paper No 1, Edmonton: University of Alberta, see: natureneedshalf.org/news-item-4/ (accessed 3 July 2014).

Solow, R. (1974) 'The economics of resources or the resources of economics', *American Economics Review Proceedings*, vol. 64, pp. 1–14.

Soulé, M. E. and Wilcox, B. A. (1980) 'Conservation biology: its scope and its challenge', in *Conservation Biology: An Evolutionary-Ecological Perspective*, eds. M. Soulé and B. Wilcox, Sunderland, MA: Sinauer.

Sukhdev, P. (2010) 'Preface' to *The Economics of Ecosystems and Biodiversity: Ecological and Economic Foundations*, ed. P. Kumar, London: Earthscan.

Suzuki, D. and McConnell, A. (1999) *The Sacred Balance: Rediscovering Our Place in Nature*, Sydney: Allen and Unwin.

Thomas, C., Cameron, A., Green, R., Bakkenes, M., Beaumont, L., Collingham, Y.,

Erasmus B., Siqueira, M., Grainger, A., Hannah, L., Hughes, L., Huntley, B., Jaarsveld, A., Midgley, G., Miles, L., Ortega-Huerta, M., Peterson, A., Phillips, O. and Williams, S. (2004) 'Extinction risk from climate change', *Nature*, vol. 427, pp. 145–148.

Tilman, D. (2012) 'Biodiversity and environmental sustainability amid human domination of global ecosystems', *Daedalis, the Journal of the American Academy of Sciences*, vol. 141, pp. 108–120.

Turner, G. (2008) 'A comparison of the *Limits to Growth* with 30 years of reality', *Global Environmental Change*, vol. 18, no. 3, pp. 397–411.

UCS (1992) 'World scientists' "Warning to Humanity"', Union of Concerned Scientists, online, available at: www.ucsusa.org/about/1992-world-scientists.html (accessed 5 August 2014).

UNEP (2011) *Recycling Rates of Metals: A Status Report*, Nairobi: United Nations Environment Programme.

Vitousek, P., Ehrlich, A. and Matson, P. (1986) 'Human appropriation of the products of photosynthesis', *BioScience*, vol. 36, no. 6, pp. 368–373.

Vitousek, P., Mooney, H., Lubchenco, J. and Melillo, J. (1997) 'Human domination of Earth's ecosystems', *Science*, vol. 277, pp. 494–499.

Wackernagel, M. and Rees, W. (1996) *Our Ecological Footprint: Reducing Human Impact on the Earth*, Gabriola Island, BC, and Canada: New Society Publishers.

Walker, B. (1995) 'Conserving biological diversity through ecosystem resilience', *Conservation Biology*, vol. 9, no. 4, pp. 747–752.

Washington, H. (2006) 'The wilderness knot'. PhD Thesis, Sydney: University of Western Sydney, online, available at: http://arrow.uws.edu.au:8080/vital/access/manager/Repository/uws:44 (accessed 2 December 2014).

Washington, H. (2013a) *Poems from the Centre of the World*, online, available at: www.lulu.com/au/en/shop/haydn-washington/poems-from-the-centre-of-the-world/paperback/product-21255751.html (accessed 2 December 2014).

Washington, H. (2013b) *Human Dependence on Nature: How to Help Solve the Environmental Crisis*, London: Earthscan.

Washington, H. and Cook, J. (2011) *Climate Change Denial: Heads in the Sand*, London: Earthscan.

Weizsäcker, E. von, Hargroves, K., Smith, M., Desha, C. and Stasinopoulos, P. (2009) *Factor 5: Transforming the Global Economy through 80% Increase in Resource Productivity*, London: Earthscan.

WCS (1980) *World Conservation Strategy: Living Resource Conservation for Sustainable Development*, IUCN/UNEP/WWF.

Whitehead, A. (1929) *Process and Reality*, New York: Harper Brothers.

Wijkman, A. and Rockstrom, J. (2012) *Bankrupting Nature: Denying our Planetary Boundaries*, London: Routledge.

Wilson, E. O. (2003) *The Future of Life*, New York: Vintage Books.

Worster, D. (1994) *Nature's Economy: a History of Ecological Ideas*, Cambridge: Cambridge University Press.

WWF (2012) Living Planet Report. World Wide Fund for Nature, online, available at: http://d2ouvy59p0dg6k.cloudfront.net/downloads/lpr_2012_summary_booklet_final.pdf (accessed 21 July 2014).

6
SOCIAL SUSTAINABILITY – UTOPIAN DREAM OR PRACTICAL PATH TO CHANGE?

> '... there is nothing more difficult to carry out, nor more doubtful of success, nor more dangerous to handle, than to initiate a new order of things.'
>
> (Niccolò Machiavelli 1532)

Introduction

What *is* a sustainable society? Sustainable for whom or for what? Is it utopia? Is it just survival over time? Many societies were stable up to the point of collapse, so this doesn't mean a society can continue indefinitely. Or is it a measure of the way society lives within its environment over time? Anand and Sen (1996) argue that social sustainability is made up of equity, diversity, social cohesions, quality of life, democracy and governance, and 'maturity' (where individuals accept responsibility for social improvement). It may seem a paradox that at the pinnacle of human material and technical achievement, we now have little or no 'community life'. We in the West find ourselves anxiety-ridden, prone to depression, worried about how others see us, and driven to consume. Modern society is not sustainable, it is *broken*, and many of us suspect this privately. Many people actually have a strong commitment to 'social fairness', but remain silent, as they think (mistakenly) that most others don't share this. The contrast between material success and social failure of many rich countries suggests a need to shift attention *away* from material goods and growth to ways of improving psychological and social well-being (Wilkinson and Pickett 2010).

We discuss here utopia, equity and equality, and justice. Social sustainability is often equated with justice and equity. These are indeed part of it, but social sustainability should be larger than these. I admit to being torn different ways when grappling with the question of social sustainability. Justice, fairness and compassion are part of my own personal ethics. However, for the last 200 years, humanity has become solipsistic and self-obsessed. Anthropocentrism or 'human-centeredness' has become the hallmark of Western society (Godfrey-Smith 1979; Smith 1998). Accordingly, when we speak of justice, fairness and compassion – they tend to be just for us, our own species. In mainstream Western society (now globalised around most of the world), we rarely apply justice and compassion *outside* our species to the rest of Nature.

We in the West have turned Nature just into a 'thing' to be used (see Crist 2012 and Chapter 8). That means that much of society (even the intelligentsia) focus only on human society and justice for humans. They don't extend the 'moral circle' (Singer 1981) *beyond our species*. Many people don't understand either ecology or environmental ethics. Most don't argue for justice for Nature. Many of those who profess a commitment to social justice at the same time tend to dismiss the intrinsic value of Nature or the 'Rights of Nature' (Washington 2006, 2013). They remain firmly rooted in anthropocentrism. Perhaps nothing illustrates this as perfectly as that the term 'environmental justice' doesn't actually mean justice for Nature *itself*, but is justice for people (usually the poor) regarding environmental problems such as pollution (Harding *et al*. 2009). Baxter (2005) and Rolston (2012) accordingly had to define 'justice for Nature' as ecological justice or *eco-justice*. Even scholars calling for a justice that considers the environmental crisis (e.g. Sachs 2002) argue that the essence of sustainability that developed from 'Our Common Future' (WCED 1987) was found as a relationship between *people and other people*, rather than people and Nature.

Now this blindness to the value of Nature concerns me deeply. The problem is one of a balance in values. Is social sustainability a key part of sustainability overall? Clearly it must be. Can we ignore the importance of ecological sustainability or an economic sustainability that controls the growth economy? No, we cannot, for humans are dependent on Nature (Washington 2013). Yet these are indeed often ignored by many social activists as irrelevant or of peripheral interest. Many of them have blinkered vision and ignore that there *is* actually an environmental crisis, as they only see the social crisis. As an environmental scientist I see both, so argue here for *both* social justice and eco-justice. In the end we must have justice for both – or we will gain neither.

Utopia

'Utopia' is said to be a 'perfect society'. It has been used to describe societies both actual and fictional. Its opposite is 'dystopia', a community or society that is in some important way undesirable or frightening. The first use of the idea of utopia was in Plato's 'Republic' (Nails 2002), which proposed a rigid class structure where 'golden' citizens were benign oligarchs who supposedly eliminated poverty through fair distribution. The word itself was coined by More (1516) in 'Utopia', a fictional Atlantic island society. Some suggest it was satirical, for 'utopia' is comprised of *ou*- meaning 'no', and *topos*, meaning 'place' so it strictly means 'no place'. However, the prefix *eu*-, meaning 'good' sounds the same, and hence it is 'eutopia' that actually means a 'good place'.

A 'Utopian socialist' movement arose in the early nineteenth century, espousing an egalitarian distribution of goods, where citizens only did enjoyable work for the common good, and had ample time for the arts and sciences. One example is Bellamy's (1888) 'Looking Backward'. Conservative or capitalist utopias have also been imagined, such as Rand's (1957) 'Atlas Shrugged'. These propose an unconstrained free market, where private enterprise allegedly brings progress (Folbre 2011). De Geus (1999) thought 'ecological utopias' could inspire green political movements. Some suggest that 'polyculturalism' and a participatory society may help us reach utopia (Spannos 2008). Some communities have been established in the hope of creating a better way of living together. Many failed, though some are growing, such as 'Twelve Tribes' (see website, online, available at: http://twelvetribes.org/). In many societies and religions, there is some myth (or perhaps cultural memory) of when humanity lived in a simple happy state of harmony with Nature (see Chapter 1). This has been called the 'Golden Age', 'Arcadia', and the 'Garden of Eden'. Anthropologists show that hunter-gatherers were the original and *most balanced* society (Gowdy 1997; Sahlins 2009), that sustainable societies existed for around two million years (Gowdy 2014, p. 34). Whether this is seen as 'utopia' will depend on one's worldview (see Chapter 8). For the techno-centrists, a hunter-gatherer society will likely be seen as anathema. They often quote Hobbes (1651) who argued that the life of humans in Nature was 'solitary, poor, nasty, brutish, and short'. Anthropology however shows this to be completely mistaken (Gowdy 1997). Those coming from an eco-centrist worldview are likely to see indigenous societies as being far better than what we have now.

The role of technology in reaching utopia is hotly debated. Buckminster Fuller (1969) presented a theory for technological utopianism. Schumacher (1973) argued for 'appropriate' technology that improved humanity (see Chapter 10). Others argue that advanced technology will cause environmental disaster or even humanity's extinction (Ellul 1964; Jensen 2006; Zerzan 2008). There are many examples of techno-dystopias portrayed in literature, such as the classics 'Brave New World' (Huxley 1932) and 'Nineteen Eighty-Four' (Orwell 1949). Currently, techno-centrists seem to think we are all heading towards a techno-utopia, while many ecocentric scholars argue that our society is dystopian in the way it treats the natural world (Berry 1999; Collins 2010). So the question remains as to whether humanity *can* ever reach utopia? Certainly with numbers over seven billion, we cannot return to being hunter-gatherers, for our population is at least 1,200 times too large (Boyden 2004). Perhaps the idea of a 'perfect society' is more an 'aspirational goal'? We should always be seeking for a 'better' society. Certainly, we can improve our social sustainability far beyond today's broken society, and move to a truly *better place*.

What is fair? Equity and equality

A key part of living sustainably into the future is about 'fairness'. The words 'equity' and 'equality' are much discussed. 'Equity' is the quality of being fair and impartial.

'Equality' is the state of being equal, especially in status, rights, income and opportunities. As we shall see, we need more of both. Widening income gaps and accelerated ecological change suggests that the mainstream global community still pays little more than lip-service to the sustainability ideal (Moore and Rees 2013, p. 49). Another way to think about it is in terms of 'sharing'. How should we live? Should just a few have most of the land, food and wealth? Should some grow fat, while others starve? Or should we share things so that everyone has a decent quality of life? Virtually every religious leader in the world has argued for compassion, for caring, sharing and *giving*. Philosopher Kahlil Gibran (1923) expressed this beautifully:

> You often say, 'I would give, but only to the deserving'.
> The trees in your orchard say not so, nor the flocks in your pasture.
> They give that they may live, for to withhold is to perish.
> Surely he who is worthy to receive his days and his nights, is worthy of all else from you?

Almost all tribal cultures had redistribution of wealth, from Chinook 'potlatch' to Mosaic 'year of jubilee'. The point is that both conservation of resources and redistribution of wealth are essential for sustainability (Czech 2013, p. 150). Ethically, there is no debate, and most people profess to agree with the idea of sharing, though many slip into denial. Wallich (1972) of the US Federal Reserve argued: 'growth is a substitute for equality of income. So long as there is growth there is hope, and that makes large income differential tolerable'. It suggests all we have to do is bake a bigger cake. Chapter 5 has shown that we *can't* in fact bake a bigger cake, when the 'oven' (the ecosystems of the world) is a fixed size, and actually degrading. We need to be fairer in how we distribute the cake we have now. Yet currently we fail to do so, and this is worsening.

The modern world has not brought greater fairness as promised, quite the opposite: 16 per cent of people live in the developed world, yet account for 78 per cent of global consumption expenditure. Meanwhile, 40 per cent of the world's population struggles to subsist on less than $2 a day (Dietz and O'Neill 2013). The richest 20 per cent lay claim to 85 per cent of the world's timber, 75 per cent of its metals and 70 per cent of its energy (UNRISD 1995). We are becoming *less fair*, not more. The average member of the top 1 per cent of the world's population is almost 2,000 times richer than someone from the poorer half (Renner 2012). Rather than ending poverty, modernism has led to the environmental crisis, and this is now impacting hugely on the poor, who rely on free ecosystem services to survive. Inequity in income distribution has increased within OECD countries, but also in major emerging economies such as China and India (OECD 2011). Growth is continuing, but the poor get *less and less of the benefits* (Layard 2005). For every $100 of global economic growth between 1990 and 2001, only 60 cents went to people with incomes of less than $1 a day (Dietz and O'Neill 2013). The mantra of development, growth and 'trickle down' is thus not actually helping the poor.

Someone is profiting from economic growth, but it's not the poor. Kopnina and Blewitt (2015) note that the 'trickle-down effect' is a myth, as much of the 'wealth' that was created in the good times was not real – just numbers on a screen – and it is the rich rather than the poor who have benefited. Sachs (2002) explains that in the neoliberal ideological view, greater 'equity' (if it matters at all) is seen as expanding the reign of the economic law of 'demand and supply' across the globe.

In their illuminating book 'The Spirit Level', Wilkinson and Pickett (2010) explain that 'inequality is bad for everyone'. They summarise extensive research showing that inequality of income worsens a whole range of social aspects. These include: health; mental illness; violence; homicides; obesity; drug use; life expectancy; competitive consumption; trust; children's educational performance; teenage births; imprisonment rates and social mobility. They develop an index of health and social problems, shown in Figure 6.1. All these social indicators are *worse* in societies with higher income inequality. Why? Because humanity is an intensely social species, so what matters is where we stand in relation to others in our own society. Human beings are driven to preserve the 'social self', and are vigilant to threats that may jeopardise social esteem or status. Increased inequality ups the stakes in the competition for status, so then status matters even more. For a species that thrives on friendship, and enjoys cooperation and trust (and which has a strong sense of fairness), it is clear that social structures based on inequality, inferiority and social

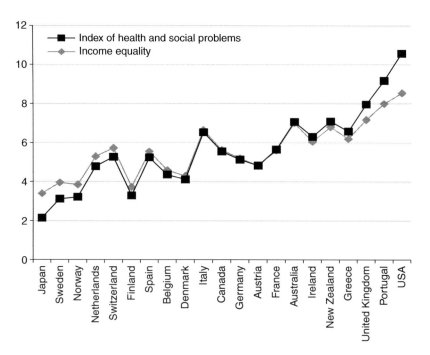

FIGURE 6.1 Index of health and social problems vs income inequality in world nations, drawn based on data from Wilkinson and Pickett (2010) *The Spirit Level: Why Equality is Better for Everyone.*

exclusion must inflict a great deal of social pain (Wilkinson and Pickett 2010, p. 216). Inequality erodes empathy, hardens hearts, undermines public trust, incites violence, saps our imagination and destroys public spirit that upholds democracy (Orr 2013, p. 288). As a result, modern society is *broken*, and likely to become increasingly dysfunctional.

Too often consumerism is regarded as if it reflected a 'fundamental human nature' of material self-interest and possessiveness. This could hardly be further from the truth. Our almost neurotic need to shop and consume is instead a reflection of how 'deeply social' we are. Living in unequal and individualistic societies, we use possessions to show ourselves in a good light and avoid appearing incompetent and inadequate (Wilkinson and Pickett 2010, p. 230). If an important part of consumerism is driven by status competition, this explains why we continue to pursue economic growth despite its apparent lack of benefits. The growth of consumerism and the weakening of community life are related (ibid.). We allow a small social elite to hoard a greater percentage of resources. The 'Gini' coefficient measures the distribution of wealth, where 0 is where everyone has an equal share and 100 is where one person has all the wealth. The Gini coefficient has been rising rapidly recently in many countries, where the United States and China are both now 47 (Maxton 2011, p. 66). It has fallen in some countries with more inclusive social policies, such as Norway and France. Such a rising trend in inequality cannot continue forever (and has not historically). A final point to consider is about how far one extends the idea of 'fairness'? Is this limited just to 'our group', is it to the whole of society, or should we extend the idea of fairness to future generations, and to all of Nature also?

What is just – and for whom?

Justice for whom? Is it only for those alive today, or does it consider our children's children? What about the ethic of the Iroquois Federation in Canada, where they believe one should plan how decisions would affect people 'seven generations hence' (Lyons 1980)? A necessary part of social sustainability has always been stated to be social justice. Why? Because if we are not just, eventually social cohesion will collapse, along finally with society itself. Sometimes this may be by revolution and war, other times by famine and plague or other environmental disasters caused by degrading the ecosystems on which we depend (Diamond 2005).

Abraham Lincoln pointed out that morally nobody is free in an unjust system (Robinson 2013). 'Justice' is much discussed by all. Sachs (2002) explains it has been firmly wedded to the idea of 'development' in post- Second World War politics, which became a 'supreme aspiration' that bound the North and South together. This mantra promised 'justice without redistribution', and thus the endless growth mantra sought to sidestep the hard questions of justice. Delinking the aspiration for justice from conventional development is thus vital for rescuing the ideal of 'justice', but also to invent social improvement that doesn't overstep ecological limits. In future, the developed world will have to accept that justice is about 'learning to take less', rather than how to give more. Anyone calling for equity will now have to speak about 'sufficiency' also (Sachs 2002). A sustainable society will require equity, greater equality and justice (locally and globally), within this generation and between our generation and future generations. A 'sustainable society', however, cannot purely focus on itself. It also needs to consider 'eco-justice' (Rolston 2012). The issue of holistic justice and ethics in a sustainable society thus firmly raises the question of a society's worldview (see Chapter 8).

Social cohesion and capital – keeping it 'together' to *act*

If we are to achieve social sustainability, then society has to function cohesively to solve problems. We need social *cohesion*, social *inclusion* and social *capital*. Social 'cohesion' describes social connectedness, including family and community well-being (Hulse 2005). It is developed through the nature and quality of interrelationships between people. It can be said to manifest in the 'sense of community'. Global events, and our increasing interconnectedness, provide new challenges. Societies may lose cohesion if people don't feel 'included'. 'Social inclusion' is the provision of certain rights to all individuals and groups, such as employment, adequate housing, health care, education and training. If they feel *included* then they feel part of a community and social cohesion is greater. Such ideas are not new, for Adam Smith (1776) believed an economy required a 'social bond' that brought social stability.

Social capital is the expected benefits derived from the cooperation between individuals and groups. The core idea is that social networks *have value*. Robert Putnam (2000) says it 'refers to the collective value of all "social networks" and the inclinations that arise from these networks to do things for each other'. Fukuyama (2002) defines it as 'shared norms or values that promote social cooperation'. Social capital is seen as a key component to maintaining democracy. It is a producer of 'civic engagement' a broad societal measure of communal health (Alessandrini 2002). Putnam believes social capital is declining in the United States, demonstrated by less trust in government and less civic participation. Television and urban sprawl play a significant role in making the United States less 'connected'. Social capital can be measured by the amount of trust and 'reciprocity' in a community. The Social Capital Foundation (SCF 2013) urges that social capital should not be confused with its manifestations. They argue it is properly the disposition to create, maintain and develop social networks, and it is a collective mental disposition, close to the 'spirit of community'. Civic engagement is thus a manifestation of social capital. Robison *et al.* (2002) propose that social capital be defined as 'sympathy', where those who have sympathy for others provide social capital.

Reaching consensus on environmental issues implies 'shared interest' and agreement among stakeholders to allow collective action. 'Consensus building' is seen as a direct positive indicator of social capital (Arefi 2003). Collective action (as in solving the environmental crisis) can thus be said to be an indicator of increasing social capital. Given that activism is a key part of solving the environmental crisis, the status of social capital within society is critical. Reaching sustainability will mean that enough people succeeded in getting their community and governments to act. There is an irony that at a time of deepening political and social fragmentation, the environmental and climate crises are demanding coordinated planetary-scale governance and management (Hulme 2009). Hence we need to build our social capital, make sure that people feel socially included, and so increase social cohesion. To reach social sustainability, we actually need a large and growing social movement that takes up the 'Great Work' (Berry 1999) of healing the Earth.

Democracy

One of the most striking characteristics of today is the 'triviality, narrowness and factual inaccuracy of our political conversations' (Orr 2013, p. 291). Social sustainability and social capital are central to democracy, and getting governments to act (Putnam 2000). Without government support and regulation, sustainability will face huge obstacles. The most common form of government is 'liberal democracy', where the majority of people elect a government. Reaching a sustainable future will require 'hard decisions', and this requires social cohesion to push governments to act. In Chapter 9 we discuss denial, and Rees (2010) notes that politicians may be strongly psychologically hard-wired to ideologies and more than usually enslaved to what he calls 'brainstem-based survival instincts', particularly the deep seated need to retain their wealth, prestige and power. Hence why public involvement for sustainability in democracy is crucial. Many scholars agree that the key problem in reaching sustainability is *lack of political will*. Any government that takes hard decisions without broad support is likely to be voted out. Many of us shy away from politics, but this is just abandoning responsibility to do something about the world's problems. If we all do this then nothing will change. 'Business-as-usual' would continue and the world would move that much closer to ecological and social disaster. Disdain for politics is in fact really 'giving up'. Many politicians would like to act on the environmental crisis, but *need to know* they have community support. Hence why social cohesion and a functioning democracy are important. Increasing social capital and cohesion makes democracy work properly for a sustainable future.

Governance

Any chance of coming through the trials of the environment crisis in a nuclear-armed world with ten billion people by 2100 will require we reckon with the thorny issues of politics and governance with wisdom, boldness and creativity (Orr 2013, p. 291). Orr (2013, p. 287) discusses governance for sustainability and notes that democracies are prone to 'spoiled child psychology', which involves a contempt for realities. This has exploded since the 1940s into a full blown 'epidemic of narcissism' in Western society. Orr observes that such failures of personality, judgment and character could multiply under the stresses likely in the 'long emergency'.

Governance is much discussed (if often not defined) in sustainability literature. Probably the most comprehensive discussion of its importance to sustainability is in Soskolne's (2008) 'Sustaining Life on Earth' and the 2014 State of the World Report (Mastny 2014).

What is 'governance'? In its simplest form it can be seen as just 'administration', while the World Bank defines it as 'the exercise of political authority and the use of institutional resources to manage society's problems and affairs'. Governance is a tool for social administration, and is both a structure and a process (Rainham *et al.* 2008, p. 173). Lafferty (2004) proposes the notion of 'governance' as the complex mix of mechanisms and instruments designed to alter, channel and influence social change. Seyle and King (2014, p. 21) describe it as the mechanisms and processes that humans use to manage their 'social, political and economic relationships with each other and the ecosphere'. As a structure, governance provides a framework for negotiation. As a process it denotes a combination of approaches used to maintain order. 'Good governance' is cultivated by the World Bank Institute research programme on 'governance and anti-corruption' (Kaufmann *et al.* 2003). It is worth noting also that often governance is confused with worldview and ethics (see Chapter 8).

Rainham *et al.* (2008, p. 173) argue there is currently a fundamental imbalance in pluralistic governance leading to:

1   obsessive focus on growth in productivity and irrational use of non-renewable resources;
2   disinvestment from public infrastructure and changes in social policy to maintain class structures;
3   little consideration of the ecological and social *consequences* associated with the processes of production and conspicuous consumption.

Soskolne *et al.* (2008, p. 3) state that current governance is incapable of arresting declines in the human condition and provides little hope of sustainability, owing to the exclusion of ecological justice (Baxter 2005). To achieve ecological justice they believe a 'true covenant' is needed as a sacred agreement. Engel (2008, p. 29) describes a 'covenant' as a time-honoured form of social relation that has served human societies in times of constitutional crisis. This can provide the moral, political and spiritual foundation for democratic and ecological global governance (Held 2004). The Earth Charter (see Chapter 2) is the ideal universal 'covenant of covenants' (Engel 2008). Bosselman (2008, p. 9) believes we must link concern for ecological integrity with concern for appropriate governance through the adoption of the Earth Charter. The charter addresses both the ethical principles of caring for ecological integrity, and the legal principles of governance. If ecological 'disintegrity' and unsustainability are the logical consequence of liberal-pluralist market societies, then any movement toward ecological sustainability must also scrutinise the dominant character of governance in these societies. Governance for sustainability is desirable, since what currently constitutes 'good governance' fails to acknowledge the primacy of natural capital and ecosystem services for sustainability (Rainham *et al.* 2008, p. 174).

There is nothing faulty with the concept of 'good governance' except for a failure to acknowledge *purpose* – governance to what end? Governance for sustainability is a concept akin to good governance, but with the 'primary object' of maintaining ecological integrity. If ecological sustainability must take precedence for the survival of our species, then we must be willing to explore potential restrictions to the forms of democracy and governance that dominate highly consumptive societies. Any form of sustainable governance will have to acknowledge the primacy of ecological integrity and equality if permanence for humanity is the goal (Rainham *et al.* 2008, p. 189). Good governance today requires stewardship of planetary life support systems, accountability, informed public opinion and sustainable conceptions of economic 'prosperity' (Hempel 2014, p. 49). If the Earth Charter was embraced by all it could play a significant role in advancing global governance toward sustaining life on Earth. It promotes a global moral community, and makes an explicit attempt to link personal ethics with public ethics (Mackey 2008, p. 197).

Governance seems at times to be used as something of a 'magic bullet' that will solve all our problems. Some even suggest it should be considered a fourth strand of sustainability. However, more correctly it can be seen as a process operating within each of the three key strands. Many who use the term seem confused, and actually seem to be talking about worldview, ethics and values (see Chapter 8). It is these that are actually the critical things we need to change. If these were changed, then I believe governance would follow. Do we need sustainability governance? Yes of course. Will we get it without changing our worldview and ethics? I doubt we will. Having said that, probably the most important aspects of governance are making it 'participatory', where the public is fully involved. This can also be called 'collaborative' or 'deliberative civic engagement' (DCE) (Prugh 2014). Deliberative governance can aid sustainability in terms of articulating purpose, framing issues, coordinating policies and organisations, legitimising processes in the eyes of the public, and mobilising action for change (Vigar and Healey 2002). However, deliberative democracy is only really successful when governments *agree to be bound* by the decision of the deliberative process. It has been suggested that we need to rebuild democracy from the bottom up, such as through a national system of neighbourhood assemblies (Orr 2013, p. 284).

We also face a problem of the 'politics of never getting there' where sustainability becomes a mirage we never reach (Foster 2008). Governments endlessly talk about sustainability and 'knowing the road', but much of it is window-dressing. The above shows we need to reflect on our current inadequacies, and develop fresh approaches and governance (Harding *et al.* 2009, p. 49). One idea is 'reflexive governance' that recognises the dynamic political and social world we live in (Voß *et al.* 2006). It requires that actors reflect on how their structures contribute to persistent problems and consider ways to fundamentally transform existing practices. Another approach is 'flexible governance', aiming to adapt, improvise and change directions (Martin 2013, p. 270). A cynic might suggest that governments will never change direction. However, 'Transition Management' in the Netherlands is a policy framework for reforming large socio-technological systems. It envisions multiple sustainable futures (Loorbach and Rotmans 2006), but also implements concrete projects. Numerous criticisms of transition management have been made (Harding *et al.* 2009, p. 50), but it represents one attempt to implement the sustainability governance we must have if we are to reach social sustainability.

So the interest in 'governance' reflects the realisation that we need to vastly improve how we respond through our institutions and governments to the environmental crisis. Jim MacNeill (2006) was a lead author of 'Our Common Future' (WCED 1987), and he reminds us that the 'forgotten imperative' of that document was that 'we could and would change the way we make decisions'. Through governance today, we are still trying to do this.

War and conflict

War and conflict are barriers to sustainability, for a society at war focuses on that war. Environmental stress or catastrophe decreases food security, and this has led to past conflicts. 'People always raid before they starve' is an observation by Dwyer (2011). Shiva (2002) argues that already we are starting to fight over access to water, and that this will increase. The climate crisis means that already many 'displacees' or 'climigrants' are looking for places to live, and the number will escalate (Brown 2011). Extreme weather and the pressure of climate change on agriculture will also stress society (Brown 2011). Uncontrolled climate change is thus likely to lead to conflict (Pauchari 2007). Civil unrest and revolution are not conducive to solving sustainability issues. Rather than unify society they fragment it, they destroy the very social cohesion we need to solve key problems. People are then 'too busy fighting the war' to think about the bigger picture, even if that might be the cause of the war. Escalating conflict due to environmental crises will make it that much harder to reach a society that acts on solutions.

The *practicality* of social sustainability

Many scholars speak of the 'essential nature' of social sustainability, especially of social equity and social justice. It is often presented as a 'given truth'. Yet it is not 'obvious', or rather, it may be obvious ethically, but it is not always obvious why it is essential *for practical reasons*. Why then are social equity and justice essential practically? The answer lies not just in ethics but in how we *deal with change*, how we face up to difficult issues and take action to solve them. It is about how a society works effectively to act. We as a society must act for sustainability. We did it with slavery, we did it with women's suffrage, now we must do so to reach a sustainable future. That means a future that is ecologically sustainable, and a future that is economically sustainable in terms of how we manage our home within ecological limits. Yet it also means a future that is socially sustainable, to the extent we can actually do the 'Great Work' of becoming sustainable.

So we come back to equity, equality and justice. The lack of each makes poverty worse, since then we do not distribute fairly what we have. Some would argue that society has never had much of these. They believe that 'human nature' is purely selfish and greedy. They argue that society has always operated on Tennyson's idea (1849) of a 'nature red in tooth and claw'. However, the idea of 'human nature' is actually dependent on your worldview. Maslow (1971) notes people ask 'how good a society does human nature permit', but he replies 'how good a human nature does society permit?'. People tend to be more altruistic than the economic model predicts (Brondizio et al. 2010). Emotions such as compassion, empathy, love and altruism are key components of the human behaviour repertoire (Manner and Gowdy 2010). Rather than assuming we are stuck with self-interested consumerism and materialism, we need to recognise that these are not fixed expressions of human nature. Instead, they reflect the characteristics of our societies (Wilkinson and Pickett 2010). Modernism teaches competition, not *cooperation*. Modernists argue that modern technology (under democracy) has brought greater equity and justice, that it has led to a better world. But has it really? The answer is a definite 'no'. Modernism and consumerism operate from the worldview of *Homo economicus* (Daly and Cobb 1994), seeing humanity primarily as consumers. It operates on the idea of 'more and more', there is never enough, and there never *can* be 'enough' (Dietz and O'Neill 2013). It idolises the greedy who have hoarded more and more resources just for them. It is not based on respect and caring, either for the Nature we are a part of, or for our fellow man. It is not a 'good' way to live, it has led directly to our broken society. The environmental crisis shows us it is also not a *sustainable* way to live. Rees (2010) concludes that under these circumstances, the long-term evolutionary selective advantage is for genes (and 'memes', see Chapter 8) that 'reinforce cooperative behaviour, even mutual altruism'. Indeed he notes that reaching sustainability will require unprecedented levels of international cooperation in the service of the common good.

Fairness and justice are ethical concepts, they are about what is 'right'. They are based on caring and compassion and even (dare I say it in today's world) on *love*. These principles have been taught by every great spiritual teacher. They are part of 'being', not of 'having' (Fromm 1979). They are not part of wanting 'more'. They don't increase the GDP, they don't help the growth economy. As such they are deemed irrelevant to neoclassical economists and the religion of 'endless growth'. Modernism and consumerism are not only anti-ethics but also anti-spiritual (Tacey 2000). If something gets in the way of growth, then it must be repressed, as it has been; hence the ignoring of ethics and the attack by modern society on a spiritual connection to Nature. The problem with this is that a spiritual connection to the Universe, and our belief in fairness and justice – are what gives life *meaning* (Berry 1988). They are a key part of what it means to be human and to live a fulfilling life. As Western society tends to dismiss fairness and justice as peripheral issues, is it any wonder that they have suffered? Equity and justice are not getting better, they are getting worse under modernism and the free market mantra of neoliberalism.

A new global solidarity for sustainability is needed, where nobody is left behind. Environmental sustainability is ultimately impossible *without* social equity, where the rich reduce their draw on materials and goods in absolute terms (Renner 2012, p. 18). Up till now people have been able to put their hope in growth. The 'trickle down' approach epitomised this fallacious idea. However, this was actually denial of the root cause: exponential growth on a finite planet that has exceeded ecological limits. We made 'endless growth' our god, not fairness and justice. The idea that justice may come from an endlessly expanding economy is no longer sustainable. People are starting to see that the emperor has no clothes. And for the 75 per cent of the world that has very little, that 'loss of hope' is not going to be acceptable. They are going to see the 25 per cent 'that have much' for what they are – greedy, selfish, and uncaring for the rest of the world, let alone future generations. Anger and desperation will result, and this will lead to a less stable world, not a more sustainable one. Unless we mend our broken society.

So it is essential we achieve better equity, equality and justice as part of social sustainability. Not just for ethical reasons (though these should be enough), but for 'practical' reasons. In the end, *inequality is bad for everyone*. Wilkinson and Pickett (2010) show the evidence is clear, that for very practical reasons of reaching a workable society, equality is better for everyone. We may indeed know the solutions needed to solve the environmental crisis (Chapter 11). However, if we don't have a society with equity (within and between generations and between humans and Nature) and justice (to all people as well as Nature) then we won't have the shared vision and social cohesion to get there. Social sustainability is thus not some 'pie in the sky' idea about reaching utopia. It is about having a society that has the social cohesion to *practically reach a sustainable future*.

What *should* social sustainability be?

Demystification requires an understanding that modern society is *broken*, with equity, equality and justice in decline. For decades, progressive politics has been weakened by the loss of any concept of a 'better society', but social sustainability now demands this. As well as potential for conflict, humans have a unique potential to be each other's best source of cooperation, learning, love and assistance. We focus on friendship and social status because the quality of social relationships has always been crucial to well-being (Wilkinson and Pickett 2010, p. 201). Our aim should be to make a more sociable society, to increase people's sense of security, to reduce fear, and offer a more fulfilling life (ibid., p. 237). We now need a popular movement capable of inspiring people with a vision of how to make our societies a better place to live. Rees (2010) concludes that the (un)sustainability crisis provides humanity with the unique privilege of writing a new, ecologically adaptive, economically viable and social equitable cultural narrative. The world is actually full of people with a stronger sense of justice and equity than we assume. People in the richest societies have been persuaded to doubt the validity and relevance of their own egalitarian values. The rise of neoliberal political and economic thinking in the 1980s and 1990s has meant that egalitarian ideas disappeared from public debate. Those with a strong sense of justice became closet egalitarians. It is time to come out of that closet and set a social course to sanity and social sustainability (Wilkinson and Pickett 2010, p. 300).

We should aim for a society that can live in a long-term balance with the world and its ecosystems. In other words, a world at peace, both amongst people and between people and Nature. Is social sustainability reaching utopia? Well, it is certainly about reaching a 'better place' ethically and practically. Let us not forget that the environmental crisis makes poverty worse, makes social equity and justice harder, and increases conflict and war. It is very much a *social crisis* too. They are entwined. Without a just, fair and cohesive community it will be difficult to reach ecological sustainability and solve the environmental crisis. Likewise, it will be harder to move to a steady state economy, in fact Wilkinson and Pickett (2010, p. 226) argue greater equality is a pre-requisite for this. We need social inclusion, we need social justice and greater equity and in fact we need a happier society – precisely because without these we will not solve the environmental crisis. This is not sentimentality but *practical reality*, and that needs to be emphasised. Social sustainability in the end is not the search for utopia, it's the adoption of a workable social structure to solve the environmental crisis, and to reach true sustainability.

More equal societies are more cohesive, have higher levels of trust, which foster 'public spiritedness'. Greater equality can help us develop the public commitment to working together we need if we are going to solve our problems. Greater equality is also key to reducing the cultural pressure to consume (see Chapters 4 and 7). Far from being 'inevitable' and unstoppable, the sense of deterioration in social well-being (and the quality of social relations in society) *are reversible* (Wilkinson and Pickett 2010, p. 33). There are many things we know we can do to become a more sustainable society. What social sustainability should be is thus a society with:

- An approach where the *highest social values* are sustainability, efficiency, sufficiency, equity and equality, beauty and community (Meadows *et al.* 2004).
- A *vision* farseeing enough, flexible enough, and wise enough not to undermine either its physical or its social systems of support (Meadows *et al.* 2004).
- A commitment to the *Lisbon Principles of Sustainable Governance*, being: responsibility; scale-matching (governing at the appropriate scale); precaution (i.e. the Precautionary Principle); adaptive management; full cost allocation (internalising externalities); and participation (Costanza *et al.* 2013).
- A focus on improving society's *quality of life* (rather than 'standard of living') through factors such as cultural expression, political freedoms and civil rights and not just increasing the GDP (Heinberg 2011), and a greater focus on 'Gross National Happiness', not GDP (GNH 2013).
- Greater commitment to *social cohesion* and social capital.
- *Greater equity* between the rich and poor nations. Rich nations should give assistance to poorer nations, but this must be for ecologically sustainable projects that improve sustainability, not promote growth.

- Greater equity and *equality of income* within nations. The most obvious way to do this is to have both minimum and maximum incomes (see Chapter 4). Governments did not actually intend or plan to lower social cohesion or to increase violence, physical and mental ill-health, or drug abuse. However, inequality of income increases all of these. It is time to tackle the core problem of inequality, for community and equality are mutually reinforcing (Wilkinson and Pickett 2010, p. 54).
- A desire to experiment with every form of *economic democracy* – employee ownership, producer and consumer cooperatives, employee representatives on boards, etc. (Wilkinson and Pickett 2010). This means a greater focus on 'cooperation' rather than competition. Cooperatives such as Mondragon in Spain emphasise this, as opposed to most corporations that focus on competition (Dietz and O'Neill 2013).
- More engagement in *environmental politics* in our democracies. Democracy needs to be widened beyond the major party system. We need a move to more participatory democracy, where the general public is more directly involved in decision-making (Elster 1998).
- Greater *discussion of worldview*, ethics and values, both human ethics and environmental ethics (Rolston 2012). We need to adopt a focus that goes back to past indigenous cultures, one of caring and sharing (see Chapter 8).
- A *shift in public values*, so that instead of inspiring 'admiration and envy', conspicuous consumption is seen as a sign of 'greed and unfairness' that damages both society and the planet (Wilkinson and Pickett 2010, p. 270).
- Greater *focus on social justice*, but also *eco-justice* for Nature, and its intrinsic value and rights. This means a greater focus on 'Earth jurisprudence' and building the 'Rights of Nature' into all nations' constitutions and legislation (Cullinan 2003, 2014).
- Improved *social inclusion*, involving fair employment, adequate housing, adequate health care, education and training.
- Greater focus on *celebrating and respecting diversity* (Braungart and McDonough 2008) and tolerance.

Many of the above have also been proposed for different reasons in other chapters. We should be doing these for multiple reasons, because they also help us to reach ecological and economic sustainability. However, while social sustainability is a 'must', as Chapter 8 discusses, we have to also consider our anthropocentric self-obsession. We need social sustainability, but not on its own. We simply cannot have it without ecological sustainability, and we cannot have it unless we also reach an economic sustainability where economy serves society, within the ecological limits of the Earth. Focusing on social sustainability alone is a myopic approach doomed to failure. Social sustainability cannot be achieved alone, just as the other two strands cannot work without social cohesion and sustainability. Many from the political Right ignore sustainability in general, while many from the Left continue to focus only on social issues. However, social sustainability must be built on the foundation of ecological sustainability, as a sustainable biosphere is non-negotiable (Rolston 2012). It is time to acknowledge the mutual interdependence involved, and to acknowledge the importance (and entwined nature) of *all three strands* of sustainability. We ignore any of them at our peril if we seek to reach a truly sustainable future.

References

Alessandrini, M. J. (2002) 'Is civil society an adequate theory?', *Third Sector Review*, vol. 8, no. 2, pp. 105–119.
Anand, S. and Sen, A. K. (1996) 'Sustainable human development: concepts and priorities', Office of Development Studies Discussion Paper, No. 1, New York: United Nations Development Programme.
Arefi, M. (2003) 'Revisiting the Los Angeles Neighborhood Initiative (LANI): lessons for planners', *Journal of Planning Education and Research*, vol. 22, no. 4, p. 384.
Baxter, B. (2005) *A Theory of Ecological Justice*, New York: Routledge.
Bellamy, E. (1888) *Looking Backward: 2000–1887*, New York: Houghton Mifflin.
Berry, T. (1988) *The Dream of the Earth*, San Francisco: Sierra Club Books.
Berry, T. (1999) *The Great Work: Our Way into the Future*, New York: Bell Tower.
Bosselman, K. (2008) 'Institutions for global governance', in *Sustaining Life on Earth: Environmental and Human Health through Global Governance*, ed. C. Soskolne, New York: Lexington Books.
Boyden, S. (2004) *The Biology of Civilisation: Understanding Human Culture as a Force in Nature*, Sydney: UNSW Press.
Braungart, M. and McDonough, W. (2008) *Cradle to Cradle: Remaking the Way we Make Things*, London: Vintage Books.
Brondizio, E., Gatzweiler, F. Zografos, C., and Kumar, M. (2010) 'The socio-cultural context of ecosystem and biodiversity valuation', in *The Economics of Ecosystems and Biodiversity: Ecological and Economic Foundations*, ed. P. Kumar, London: Earthscan.
Brown, L. (2011) *World on the Edge: How to Prevent Environmental and Economic Collapse*, New York: W. W. Norton and Co.
Collins, P. (2010) *Judgment Day: The Struggle for Life on Earth*, Sydney: UNSW Press.
Costanza, R., Alperovitz, G., Daly, H., Farley, J., Franco, C., Jackson, T., Kubiszewski, I., Schor, J. and Victor, P. (2013) 'Building a sustainable and desirable economy-in-society-in-nature', in *State of the World 2013: Is Sustainability Still Possible?*, ed. L. Starke, Washington: Island Press.
Crist, E. (2012) 'Abundant Earth and the population question', in *Life on the Brink: Environmentalists Confront Overpopulation*, eds. P. Cafaro and E. Crist, Georgia: University of Georgia Press, pp. 141–151.
Cullinan, C. (2003) *Wild Law: A Manifesto for Earth Justice*, Totnes, Devon: Green Books.
Cullinan, C. (2014) 'Governing people as members of the Earth community', in *State of the World 2014: Governing for Sustainability*, ed. L. Mastny, Washington: Island Press.
Czech, B. (2013) *Supply Shock: Economic Growth at the Crossroads and the Steady State Solution*, Canada: New Society Publishers.
Daly, H. and Cobb, J. (1994) *For the Common Good: Redirecting the Economy toward Community, the Environment, and a Sustainable Future*, Boston: Beacon Press.
Diamond, J. (2005) *Collapse: Why Societies Choose to Fail or Succeed*, New York: Viking Press.
Dietz, R. and O'Neill, D. (2013) *Enough is Enough: Building a Sustainable Economy in a World of Finite Resources*, San Francisco: Berrett-Koehler Publishers.

Dwyer, G. (2011) *Climate Wars: The Fight for Survival as the World Overheats*, Canada: Vintage Books.
Ellul, J. (1964) *The Technological Society* (Trans. John Wilkinson), New York: Knopf.
Elster, J. (ed.) (1998) *Deliberative Democracy (Cambridge Studies in the Theory of Democracy)*, Cambridge, UK: Cambridge University Press.
Engel, J. (2008) 'A covenant of covenants: a federal vision of Global 27 governance for the twenty-first century', in *Sustaining Life on Earth: Environmental and Human Health through Global Governance*, ed. C. Soskolne, New York: Lexington Books.
Folbre, N. (2011) 'Utopian capitalism', *New York Times*, 26 December 2011.
Foster, J. (2008) *The Sustainability Mirage: Illusion and Reality in the Coming War on Climate Change*, London: Earthscan.
Fromm, E. (1979) *To Have or to Be*, New York: Abacus.
Fukuyama, F. (2002) 'Social capital and development: the coming agenda', *SAIS Review*, vol. 22, no. 1, pp. 23–37.
Fuller, B. (1969) *Operating Manual for Spaceship Earth*, Carbondale, IL: Southern Illinois University Press.
Geus, M. de (1999) *Ecological Utopias: Envisioning the Sustainable Society*, Utrecht, Netherlands: International Books.
Gibran, K. (1923) *The Prophet*, New York: Alfred A. Knopf.
GNH (2013) Gross National Happiness website, online, available at: www.grossnationalhappiness.com/ (accessed 22 July 2013).
Godfrey-Smith, W. (1979) 'The value of wilderness', *Environmental Ethics*, vol. 1, pp. 309–319.
Gowdy, J. (1997) *Limited Wants, Unlimited Means: a Reader On Hunter-gatherer Economics and the Environment*, Washington: Island Press.
Gowdy, J. (2014) 'Governance, sustainability, and evolution', in *State of the World 2014: Governing for Sustainability*, ed. L. Mastny, Washington: Island Press.
Harding, R., Hendriks, C. and Faruqi, M. (2009) *Environmental Decision-making: Exploring Complexity and Context*, Sydney: Federation Press.
Heinberg, R. (2011) *The End of Growth: Adapting to Our New Economic Reality*, Canada: New Society Publishers.
Held, D. (2004) *Global Covenant: the Social Democratic Alternative to the Washington Consensus*, Cambridge, MA: Polity Press.
Hempel, M. (2014) 'Ecoliteracy: knowledge is not enough', in *State of the World 2014: Governing for Sustainability*, ed. L. Mastny, Washington: Island Press.
Hobbes, T. (1651) *Leviathan or The Matter, Forme and Power of a Common Wealth Ecclesiasticall and Civil*, London.
Hulme, M. (2009) *Why We Disagree About Climate Change: Understanding Controversy, Inaction and Opportunity*, Cambridge: Cambridge University Press.
Hulse, K. (2005) 'Housing, housing assistance and social cohesion in Australia', Monash Research Centre Final Report, online, available at: www.ahuri.edu.au/publications/p50300/ (accessed 2 December 2014).
Huxley, A. (1932) *Brave New World*, London: Chatto and Windus.
Jensen, D. (2006) *Endgame, Volume 1: The Problem of Civilization*, New York: Seven Stories Press.
Kauffmann, D., Krayy, A., and Mastruzzi, M. (2003) *Governance Matters III: Governance Indicators for 1996–2002*, Washington, DC: World Bank.
Kopnina, H. and Blewitt, J. (2015) *Sustainable Business: Key Issues*, London: Routledge.
Lafferty, W. (2004) 'Governance for sustainable development: lessons and implications', in *Governance for Sustainable Development: the Challenge of Adapting Form to Function*, ed. W. Lafferty, Cheltenham: Edward Elgar Publishing.

Layard, R. (2005) *Happiness: Lessons from a New Science*, New York: Penguin Press.
Loorbach, D. and Rotmans, J. (2006) 'Managing transitions for sustainable development', *Understanding Industrial Transformation: Views from Different Disciplines*, eds. X. Olsthoorn and A. Wieczorek, Dordrecht: Springer, pp. 187–206.
Lyons, O. (1980) 'An Iroquois perspective', in C. Vecsey and R. Venables (eds) *American Indian Environments: Ecological Issues in Native American History*, New York: Syracuse University Press, pp. 173–174.
Machiavelli, N. (1532) *The Prince*, Florence, Italy: Antonio Blado d'Asola.
Mackey, B. (2008) 'The Earth Charter, ethics and global governance', in *Sustaining Life on Earth: Environmental and Human Health through Global Governance*, ed. C. Soskolne, New York: Rowman and Littlefield Publishers.
MacNeill, J. (2006) 'The forgotten imperative of sustainable development', *Environmental Policy and Law*, vol. 36, nos 3/4, pp. 167–170.
Manner, M. and Gowdy, J. (2010) 'The evolution of social and moral behaviour: evolutionary insights for public policy', *Ecological Economics*, vol. 69, no. 4, pp. 753–761.
Martin, B. (2013) 'Effective crisis governance', in *State of the World 2014: Governing for Sustainability*, ed. L. Mastny, Washington: Island Press.
Maslow, A. (1971) *The Farthest Reaches of Human Nature*, New York: Viking Press.
Mastny, L. (ed.) (2014) *State of the World 2014: Governing for Sustainability*, Washington: Island Press.
Maxton, G. (2011) *The End of Progress: How Modern Economics has Failed Us*, West Sussex, UK: John Wiley and Sons.
Meadows, D., Randers, J. and Meadows, D. (2004) *Limits to Growth: The 30-Year Update*, Vermont: Chelsea Green.
Moore, J. and Rees, W. (2013) 'Getting to one-planet living', in *State of the World 2013: Is Sustainability Still Possible?*, ed. L. Starke, Washington: Island Press.
More, T. (1516) *Utopia: Of a Republic's Best State and of the New Island Utopia*, London: More.
Nails, D. (2002) *The People of Plato: a Prosopography of Plato and Other Socratics*, Cambridge, MA: Hackett Publishing.
OECD (2011) *Divided We Stand: Why Inequality Keeps Rising*, Organisation for Economic Development and Cooperation, Paris, online, available at: www.oecd.org/document/51/0,3746,en_2649_33933_49147827_1_1_1_1,00.html (accessed 2 February 2012).
Orr, D. (2013) 'Governance in the long emergency', in *State of the World 2013: Is Sustainability Still Possible?*, ed. L. Starke, Washington: Island Press.
Orwell, G. (1949) *Nineteen Eighty-Four*, London: Secker and Warburg.
Pauchari, R. (2007) 'Intergovernmental panel on climate change': Nobel Peace Prize 2007, acceptance speech by Dr R. Pauchari for Nobel Peace Prize, online, available at: www.webcitation.org/5nCTtMu3T (accessed 28 July 2013).
Prugh, T. (2014) 'Building a culture of engagement', in *State of the World 2014: Governing for Sustainability*, ed. L. Mastny, Washington: Island Press, pp. 250–251.
Putnam, R. (2000) *Bowling Alone: the Collapse and Revival of American Community*, New York: Simon and Schuster.
Rainham, D., McDowell, I. and Krewski, D. (2008) 'A sense of possibility: what does governance for health and ecological sustainability look like?', in *Sustaining Life on Earth: Environmental and Human Health through Global Governance*, ed. C. Soskolne, New York: Lexington Books.
Rand, A. (1957) *Atlas Shrugged*, New York: Signet.
Rees, W. (2010) 'What's blocking sustainability? Human nature, cognition and denial, *Sustainability: Science, Practice and Policy*, vol. 6, no. 2 (ejournal), online, available at: http://sspp.proquest.com/archives/vol.6iss2/1001–012.rees.html (accessed 21 June 2014).

Renner, M. (2012) 'Making the green economy work for everybody', in *Moving toward Sustainable Prosperity: State of the World 2012*, ed. L. Starke, Washington: Island Press.

Robinson, K. S. (2013) 'Is it too late?', *State of the World 2013: Is Sustainability Still Possible?*, ed. L. Starke, Washington: Island Press.

Robison, L., Schmid, A. and Siles, M. (2002) 'Is social capital really capital?', *Review of Social Economy*, vol. 60, pp. 1–21.

Rolston III, H. (2012) *A New Environmental Ethics: the Next Millennium of Life on Earth*, London: Routledge.

Sachs, W. (2002) 'Ecology, justice and the end of development', in *Environmental Justice, Discourses in International Political Economy – Energy and Environmental Policy*, vol. 8, eds J. Byrne C. Martinez and L. Glover, London UK: Transaction Publishers, pp. 19–36

Sahlins, M. (2009) 'Hunter-gatherers: insights from a golden affluent age', *Pacific Ecologist*, Winter, 2009, pp. 3–8, online, available at: http://pacificecologist.org/archive/18/pe18-hunter-gatherers.pdf (accessed 16 July 2013).

SCF (2013) 'The Social Capital Foundation', Citizendium, The Citizen Compendium, online, available at: http://en.citizendium.org/wiki/The_Social_Capital_Foundation.

Schumacher, E. (1973) *Small Is Beautiful: A Study of Economics As If People Mattered*, New York: Blond & Briggs.

Seyle, D. and King, M. (2014) 'Understanding governance', in *State of the World 2014: Governing for Sustainability*, ed. L. Mastny, Washington: Island Press.

Shiva, V. (2002) *Water Wars: Privatization, Pollution, and Profit*, Canada: Transcontinental Publishing.

Singer, P. (1981) *The Expanding Circle: Ethics and Sociobiology*, New York: Farrar, Straus & Giroux.

Smith, A. (1776) *The Wealth of Nations: an Inquiry into the Nature and Causes of the Wealth of Nations*, London: W. Strahan and T. Cadell.

Smith, M. J. (1998) *Ecologism: Towards Ecological Citizenship*, Buckingham: Open University Press.

Soskolne, C. (ed.) (2008) *Sustaining Life on Earth: Environmental and Human Health through Global Governance*, New York: Lexington Books.

Soskolne, C., Westra, L., Kotze, L. Mackey, B., Rees, W. and Westra, R. (2008) 'Global and local contexts as evidence for concern', in *Sustaining Life on Earth: Environmental and Human Health through Global Governance*, ed. C. Soskolne, New York: Lexington Books.

Spannos, C. (2008) 'What is real utopia?', *Z Magazine*. Friday, 4 July, online, available at: www.zcommunications.org/what-is-real-utopia-by-chris-spannos (accessed 8 July 2013).

Tacey, D. (2000) *Re-enchantment: the New Australian Spirituality*, Australia: Harper Collins.

Tennyson, A. (1849) 'In Memoriam A. H. H.', poem by Alfred, Lord Tennyson.

UNRISD (1995) *States of Disarray: the Social Effects of Globalisation*, Geneva: United Nations Research Institute for Social Development.

Vigar, G. and Healey, P. (2002) 'Environmentally respectful planning: five key principles', *Journal of Environmental Planning and Management*, vol. 45, no. 4, pp. 517–532.

Voß, J.-P., Bauknecht, D. and Kemp, R. (2006) *Reflexive Governance for Sustainable Development*, Cheltenham: Edward Elgar.

Wallich, H. C. (1972) 'Zero growth', *Newsweek*, 24 January.

Washington, H. (2006) 'The wilderness knot'. PhD Thesis, Sydney: University of Western Sydney, online, available at: http://arrow.uws.edu.au:8080/vital/access/manager/Repository/uws:44.

Washington, H. (2013) *Human Dependence on Nature: How to Help Solve the Environmental Crisis*, London: Earthscan.

WCED (1987) *Our Common Future*, World Commission on Environment and Development, London: Oxford University Press.
Wilkinson, R. and Pickett, K. (2010) *The Spirit Level: Why Equality is Better for Everyone*, London: Penguin Books.
Zerzan, J. (2008) *Twilight of the Machines*, Washington: Feral House.

7
OVERPOPULATION AND OVERCONSUMPTION

> *As the knowledgeable scientific community has repeatedly explained, the only safe course for humanity is to humanely end population growth as soon as possible and start a slow decline toward a sustainable number, reduce wasteful consumption, and focus on making key technologies less environmentally damaging.*
> *(In 'Sustainable Futures', Paul and Anne Ehrlich 2014)*

We have discussed what economic, ecological and social sustainability *should* be. However, again and again we have encountered the entwined problems of 'population' and 'consumption', which deserve a chapter of their own. To demystify 'sustainability', we must demystify the key drivers of unsustainability. That means we must tackle the myths around population and consumption. Actually, it is about *over*-population and *over*-consumption. In terms of solving the environmental crisis, the four biggest elephants we don't see standing in the room are the growth economy, climate change, and population and consumption. I have discussed climate change denial elsewhere (Washington and Cook 2011), while the growth economy is covered in Chapter 4. However, we cannot meaningfully discuss 'sustainability' unless we grapple with overpopulation and overconsumption. They are the key causes of our current unsustainability. They are also issues that society (and even much of academia!) fervently seek to ignore. We must take the bull by the horns and discuss why action on both is essential, indeed why it is a 'prerequisite' for sustainability.

Overpopulation

Population growth exacerbates most environmental problems, and is a key underlying part of the growth economy (Chapter 4). Apart from daring to question the growth economy, nothing else seems to raise such passion as suggesting we should 'limit our numbers'. Into it comes issues such as religion, racism, social and environmental justice, equity and poverty. There is also no more 'taboo' issue politically than population (with the exception of questioning the growth economy!). Collectively, the public and governments have been shying away from it for decades. Yet Hulme (2009) notes that if there is a 'safe' level of greenhouse gases to avoid runaway climate change, then 'is there not also a desirable world population?'. Between 1960 and 2012 the human population more than doubled from three billion to over seven billion (and the global economy went up sevenfold, Crist 2012, p. 5). In environmental science, controversy has raged as to whether consumption or population growth is most culpable. We shall see here that this is a false dilemma, for the answer is *both*. Yet in the last two decades, population has become something of a non-issue (or even a taboo word), so that a key driver of the environmental crisis is often ignored (Crist 2012).

A common knee-jerk reaction to raising the issue of population is to call those who raise it 'anti-human' or a 'racist' (Collins 2010). The concept of 'race' of course is dubious genetically (and not one I believe in). In terms of the environmental crisis, the ethnic origin of people is not the point. The total numbers and distribution of people (and their impact on the Earth's carrying capacity) *is* the point. Unsustainable population growth pushes the world beyond its carrying capacity (Catton 1982). The world is finite, and we know that human numbers have grown exponentially, so that they are now larger than ever before. Our global population is more than 7.2 billion people. Various projections indicate that by 2050 the population will grow to between 8.3 and 10.9 billion people (UN 2012), with a predicted medium figure being 9.6 billion by 2050. It would make a big difference to our sustainability prospects if it was the lowest possible figure. So the idea that the population explosion is 'over' often portrayed in the media is wrong (Campbell 2012). However, if we take action, we could in fact halt our population at eight billion (Engelman 2012a) (and then reduce it to an ecologically sustainable level).

In 1968 Paul Ehrlich published 'The Population Bomb', which alerted the world to the dangers of exponentially growing population. He was later part (Ehrlich *et al.* 1977) of coining the equation:

$$I = PAT$$

Environmental impact equals population × affluence × technology. Our impact on the Earth is thus the number of people times their affluence (per capita consumption of resources) times the technology we use. Of course, like most environmental scientists, one must accept that most of the impact from pollution and carbon emissions are *currently* coming from the consumers in the developed world (Monbiot 2009). However, the developing world is rapidly seeking to catch up. If this is done using traditional carbon-polluting industry, then the result will be steeply accelerating global carbon emissions, resource consumption, and other pollution. Indeed, this is already happening. The technology used to 'catch up' will thus be a critical factor (see Chapter 10), as will the question of whether the developing world seeks to catch up to the incredibly wasteful American (or Australian) level. However, improving technology or reducing affluence can only reduce our impact so far. In the end the numbers of people themselves count. The sheer number of consumers matters as much as the fact that many are now consuming more. A big population has a big impact, especially as the developing world expands its economy. Despite a 30 per cent increase in resource efficiency, global resource use has expanded by 50 per cent over 30 years (Flavin 2010). This is mainly due to the increasing affluence of the large populations in the developing world. This is why China is now the world's biggest carbon polluter, while India now ranks sixth (Assadourian 2010). Accordingly, we need to target *all three* components of I=PAT if we seek to reduce human impact: containing population, limiting affluence and cleaning technology. These are key tasks of any meaningful interpretation of what sustainability should be.

Why is population such a diabolical policy issue? Because it cuts at the heart of the received wisdom of two million years of human evolution, where 'more' people was always better (Washington 1991). 'More' meant we could gather more food, cut down more forest, hunt more animals, defend ourselves better, and ostensibly gather more taxes for the State (though sometimes this can mean even more money needs to be spent). 'More people' as a concept until the last 100 years has always been seen as a 'good thing' for society. Clearly people love babies, so it goes against the grain to say we should have fewer. Even authors of sustainability classics such as 'Cradle to Cradle' (Braungart and McDonough 2008, p. 5, 66) balk of talking about stabilising population. Collins (2010) believes that at the core of the population problem is a 'conflict of rights' – the right of the individual to reproduce, and the right of other species to continue to exist. It is very hard for us to understand in our hearts that now 'more' is no longer better. Add to this the religious discouragement of birth control methods (e.g. the Catholic Church). Add to that the fundamental desire of governments to have more citizens and greater power. Population ecologist Meyerson (see Meyerson 2007) explains:

> Conservatives are often against sex education, contraception and abortion and they like growth – both in population and in the economy. Liberals usually support individual human rights above all else and fear the coercion label and therefore avoid discussion of population growth and stabilisation. The combination is a tragic stalemate that leads to more population growth.

Crist (2012) points out also that environmentalists and the political Left have both blundered badly in failing to face up to population growth. In 1994 the UN 'Cairo' conference stopped talking about 'family planning' and instead spoke of 'women's reproductive health' (funding for family planning then dropped worldwide). At that time population became something of a taboo word, as it was portrayed as infringing on 'women's rights'. Many in the Left have referred to the failed forced sterilisation programme in India, suggesting (erroneously) that most family planning was coercive (Campbell 2012). In fact family planning is about giving women the choice as to when to use their 'right' to have children. In fact if family planning and contraception were made universally available, the evidence is that population growth would stabilise and then start to decline (Engelman 2012a, p. 236). Another problem has been a common (if not universal) trend in feminism and the political Left to argue against population control as coercive (Kolankiewicz and Beck 2001), though this may be starting to change (Weeden and Palomba 2012, p. 270).

Population exacerbates all other environmental problems. Butler (2012, p. 167) notes that both climate change and the extinction crisis are merely symptoms of ecological overshoot by an obese humanity. Overpopulation means cutting more forest for farmland, over-farming land so that it erodes, killing more 'bush meat' (wild animals) for food, over-fishing the rivers and seas. It means burning more fossil fuels as a way of fuelling 'development'. Many scholars write of the need for a 'smaller ecological footprint', but as Dietz and O'Neill (2013, p. 78) point out: 'we need smaller footprints, but we also need fewer feet'. And yet despite all this, talking about overpopulation is still controversial. As Chris Rapley, the director of London's Science Museum notes:

> So controversial is the subject that it has become the 'Cinderella' of the great sustainability debate – rarely visible in public, or even in private. In interdisciplinary meetings addressing how the planet functions as an integrated whole, demographers and population specialists are usually notable by their absence.
>
> *(BBC 2006)*

It is worth considering what an ecologically sustainable population number for the Earth might be. Crist (2012) points out that this question should really be what is the number (and at what level of consumption) that can live on Earth without turning it into a human colony founded on the genocide of the non-human. Biocapacity data suggest that if we made no change at all to consumption patterns, we could currently sustain a population of four to five billion. Our ecological footprint suggests no more than 4.7 billion people (Engelman 2013, p. 9), but not if every one of those lived at the US standard, where the Earth could sustain only a quarter of today's population, or 1.75 billion people (Assadourian 2013, p. 115). If everybody on Earth shared a modest standard of living, midway between the richest and the poorest, that figure would be around three billion (PM 2010). If we stabilise world population at eight billion we could reduce it to six billion by the end of the century and to a sustainable two to three billion by the end of the following century (Staples and Cafaro 2012, p. 296). Collins (2010) concludes that human population numbers are now so far beyond sustainability as to render the concept irrelevant. The world is already *over*populated. We cannot live in harmony with Nature when our numbers are degrading the world's life support systems.

To feed all the worlds people by 2050, given rising populations and incomes, food production must increase by 70 per cent according to a FAO report, but at the same time the report noted 25 per cent of the world's land is degraded and water is becoming increasingly scarce and polluted, both above and below ground (FAO 2011). It is hard to see how it is possible in the future to adequately feed nine or ten billion people, given the many accelerating and interconnected environmental problems that food production now faces (Brown 2012). It would only perhaps be possible by destroying much of our remaining natural areas and biodiversity to increase cropland by a fifth (Erb et al. 2009). Even if everyone lived at the China standard, we would need slightly more than one Earth to supply them (Moore and Rees 2013, p. 41). Engelman (2012a) argues we could stabilise world population at eight billion if we applied the strategies laid out below. Then we would need to reduce overall population over time towards a lower, ecologically sustainable number, at least half that (possibly two billion would be a more responsible and ecologically realistic figure, Ehrlich 2013). While population remains in the seven to eight billion range, greenhouse gas and material use will have to come way down from Western levels to reduce rapidly the overall impact (Engelman 2013b, p. 9).

Overpopulation *can* be tackled by nine humane (non-coercive) strategies (Engelman 2012a, p. 122):

1 Assure access to contraceptives.
2 Guarantee education through secondary school for all (with particular focus on girls).
3 Eradicate gender bias from laws, economic opportunity, health and culture.
4 Offer age-appropriate sexuality education for all.
5 End all policies that reward parents financially based on their number of children.
6 Integrate teaching about population, environment and development into all school curricula.
7 Put full pricing on environment costs and impacts.
8 Adjust to population ageing, rather than trying to delay it through government programmes aimed at boosting birth rates.
9 Convince leaders to commit to ending population growth through the exercise of human rights and human development.

Note that neither I nor other environmental scientists are suggesting compulsory sterilisation or forced euthanasia for the old. These are 'myths' promulgated by those who deny the problem, and seek to label population realists as 'anti-human'. The fact that humane and con-coercive strategies (such as the nine above) can work is attested to by the fact that Iran was able to halve its population growth rate from 1987 to 1994 (Brown 2011). Brazil, Iran, Japan, Sri Lanka and Thailand have achieved replacement fertility levels in a matter of a decade after strong government communication campaigns combined with affordable family planning (Ryerson 2012, p. 251). Population Media (see website, online, available at: www.populationmedia.org) has also had great success through education in many nations. If we are going to accept reality then we need to accept all of it. Population growth is a key contributor to the environmental crisis. Our planet is overpopulated by humans and we cannot reach sustainability with our current numbers, let alone increasing numbers. We need to stop denying this and develop rational, ethical, humane and non-coercive, ecologically-based, population policies. To deny this just exacerbates denial of the causes of the environmental crisis (O'Connor and Lines 2008). Limiting human numbers will not guarantee we reach sustainability, but Kolankiewicz (2012) observes that not doing so means certain failure. 'More people' is no longer 'better', but far worse. It keeps us from reaching a truly sustainable future.

Dealing with consumerism and overconsumption

Humanity's indigenous cultures in the past (and today still in places) had worked out how to manage the commons sustainably (Ostrom 1990). However, modern humanity has been plagued by the 'Tragedy of the Commons' (Hardin 1968), which one can describe as free access and unrestricted demand for a finite resource that ultimately depletes it. Hence modern society has overgrazed pasture, overlogged forests and overfished our fisheries. Similarly we are now mining resources and creating pollution far beyond what natural systems can detoxify. Our lifestyles and our consumption are on a violent collision course with Nature, so we cannot continue to live as we have (Wijkman and Rockstrom 2012, p. 11). What drives this irrational behaviour that ultimately decreases society's well-being? We will find here again that at the heart of this is the unsustainable idea of endless growth. Chapter 4 showed us that the growth economy is broken and fundamentally unsustainable. Had economists collapsed in deepest shame on being shown in the 1930s or the 1970s that their theories fell down against the Second Law of Thermodynamics, we would have made a great deal more progress to sustainability today (Zencey 2013, p. 83). However, they did not, and most still in fact refuse to admit that growth economics violates the basic ecological realities of the Earth. Hulme (2009) notes there is a paradox at the heart of economic analyses of climate change: 'the presumption of growth'. Yet many believe that the fixation on growth and the assumption that 'increasing consumption' is the path to well-being is precisely why we have an environmental crisis. Layard (2005) notes:

> Economic growth is indeed triumphant, but to no point. For material prosperity does not make humans happier: the 'triumph of economic growth' is not a triumph of humanity over material wants; rather it is the triumph of material wants over humanity.

The UN Department of Economic and Social Affairs notes that so many of the components of existing economic systems are locked into the use of non-green and non-sustainable technology, while simultaneously so much is at stake in terms of the high cost of moving out of them. The result is 'policy paralysis' (Moore and Rees 2013, p. 40). Chapter 4 lists the many actions we can take immediately to move to real economic sustainability. The key (and poorly discussed) problem of our growth economy is consumerism and overconsumption. After all, environmental impact comes from population *times* consumption. As Paul Ekins (1991) has noted, a *sustainable* 'consumer society' is a contradiction in terms. Even in the 1950s, retail analyst Victor Lebow (1955) concluded:

> Our enormously productive economy demands that we make consumption our way of life, that we convert the buying and use of goods into rituals, that we seek our spiritual satisfaction and our ego satisfaction in consumption.... We need things consumed, burned up, worn out, replaced and discarded at an ever-increasing rate.

For some people today, consumption has become the meaning of life, the 'chief sacred', the 'mystery before which one bows' (Ellul 1975). Yet the same consumption that has lately become the 'meaning of life' is now revealed as the greatest hazard to life. Ultimately, we cannot roll back denial of the environmental crisis unless we roll back our consumer worldview (Starke and Mastny 2010). Consumer cultures exaggerate the forces that have allowed human societies to 'outgrow' their environmental support systems (Assadourian 2010). We have not in fact outgrown the ecosystem services that support us (and simply cannot). Humanity remains completely dependent on Nature (Washington 2013). Hamilton (2010) argues that many of us have constructed a 'personal identity' through shopping and consumerism. We have substituted consumerism for *meaning* (Collins 2010). Asking people embedded in the consumer myth to change their consumerism may thus be like asking them to change their identity. The issue is further complicated in that the oft-stated aim to 'end global poverty' has become synonymous with the spread of a high consumption standard of living (Crist 2012, p. 7). However, it is time to openly discuss the urgent need to alter today's consumer consciousness. It is neither healthy, ethical, nor sustainable.

Sukhdev (2010) notes we have become attuned to giving 'yes' answers for trade-off choices that result in more consumption, more private wealth, more physical capital – as against *better* consumption, more public wealth, more natural capital. Society extends this false logic implicitly when we ignore the depletion of forests, fisheries, and so forth. This is part of a perverted logic of 'promoting growth' or 'promoting development', without defining these terms in holistic or equitable ways (from either an intra- or inter-generational perspective) (Sukhdev 2010). While we promote endless growth on a finite planet, mainstream society doesn't think about whether this is fair to future generations, or indeed fair to the rest of life we share this planet with.

The consumer ethic, seen as 'natural' by consumers, is actually a *cultural teaching*, a purposeful social construct (Assadourian 2010). Following the Second World War, the United States was 'blessed' with great industrial capacity, and large numbers of under-employed workers (returned soldiers). To take advantage of this abundant labour and break people out of their wartime habit of thriftiness, industry organised to legitimise profligate consumption, to make it a 'spiritual activity' (Rees

2008). In fact, people resisted the throwaway society when it was first promulgated, as they believed in *thriftiness*. Three sectors aided the spread of consumerism: the car industry, the pet industry and the fast food industry (Assadourian 2013, p. 117). The notion of 'perpetual growth' is thus a social construct vastly escalated as a transition strategy to reboot the economy after the Second World War. It has now (well and truly!) run its useful course. What society has constructed it can theoretically deconstruct and replace. The time has come for a new social contract that recognises humanity's collective interest in designing a better form of prosperity for a world with ecological limits (Moore and Rees 2013, p. 49). This is a challenge, but it is also an opportunity to get things right for a sustainable future.

Tacey (2000) points out that consumers in Western society are spiritually empty, so shopping temporarily fills this void (see Chapter 8). Mass consumption requires that consumer demand remains insatiable (Westra 2008). Preventing the collapse of human civilisation requires nothing less than a wholesale transformation of dominant consumer culture (Flavin 2010). Consumption has gone up sixfold since 1960, but numbers have only grown by a factor of 2.2. Consumer expenditure per person has almost tripled (Assadourian 2010). According to the Global Footprint Network, humanity now uses the resources and services of 1.5 Earths (GFN 2013), an unsustainable situation. If all the world were to adopt American (or Australian) lifestyles, we would need at least four more planets to supply them (Graff 2010). Assadourian (2010) suggests three goals to tackle consumerism. First, consumption that undermines well-being has to be discouraged. Second, we need to replace private consumption of goods with public consumption of services (e.g. libraries, public transport). Third, necessary goods must be designed to last and be 'cradle to cradle' recyclable. Wilkinson and Pickett (2010, p. 226) point out also that if we improve 'equality of income' in our societies, then consumer pressure will decline.

Rees (2010) argues that if we are to reduce the human eco-footprint:

> the fetishistic emphasis in free-market capitalist societies on individualism, competition, greed, and accumulation must be replaced by a renewed sense of community, cooperative relationships, generosity, and a sense of sufficiency; short-term material wants must give way to long-term basic needs.

Consumer cultures will have to be reengineered into a culture of sustainability, so that living sustainably as a 'citizen' feels as natural as living as a consumer does today. It will not be easy, and will be forcefully resisted by a myriad of interests, such as the fossil fuel industry, big agribusiness, food processors, the fast food industry, car manufacturers, advertisers, etc. However, either we find ways to wrestle our cultural patterns out of the grip of vested interests, or Earth's ecosystems will decline further, and bring down the consumer culture in a much crueller way. To break free of consumerism, we will need to use all our social institutions: business, media, marketing, government, education, social movements, and social traditions (Assadourian 2013). We will need to engage the public in a conversation about the growing and unsustainable costs of the consumer society (Assadourian 2013,

p. 119). 'Choice editing', or strategies for sustainability where you don't provide people with some 'high impact' products will be needed. Consumer choice editing or restriction of unsustainable products can help eliminate unsustainable choices (Kopnina and Blewitt 2015, p. 228).

The 'finger of blame' can validly be pointed directly to the main agent of the 'brown' economy: *corporations* and the rules that govern their operations (Sukhdev 2013, p. 144). Corporations are responsible for 60 per cent of world GDP (Prugh 2013, p. 111). Better policies are needed regarding accounting practice, taxation, financial leverage and advertising. These could result in a 'new corporate model', an agent for tomorrow's sustainable economy. We need:

1 tax shifting;
2 rules and limits to govern financial leverage;
3 advertising norms and standards that make it more responsible and accountable;
4 for all major corporate externalities (positive or negative) to be measured, audited and reported in annual statements (Sukhdev 2013, p. 144).

We also need to ask what drives today's unsustainable consumption? It is no exaggeration to say that corporate advertising is the biggest single force driving consumer demand today. Daly (2008) suggests an Advertising Tax to reduce advertising pressure. Victor (2008) argues advertising must now return to *strictly factual* information. Sukhdev (2013, p. 150) suggests four strategies for more accountable advertising:

1 disclose lifespan of products in all ads;
2 disclose countries of origin;
3 recommend on the product 'how to dispose of it';
4 voluntarily submit a 10 per cent development donation on total advertising spent in developing countries to support local sustainability projects.

Dematerialising our economy

Clearly, an important part of reducing humanity's impact on society is using less materials and energy. Meadows *et al.* (2004) note that the 'good news' is that current high rates of throughput are not necessary to support a decent standard of living for all the world's people. However, the bad news is that many crucial 'sources' of natural capital (such as forests, fisheries etc.) are emptying or degrading and many 'sinks' (i.e. the ability to clean up pollution) are filling up or overflowing. The reality that we need to dematerialise and use less energy is hardly novel, as in 1993 the Business Council for Sustainable Development concluded that the industrialised world would require reductions in 'material throughput, energy use and environmental degradation of over 90 per cent' by 2040 for reasons of equity and living within the planet's means (BCSD 1993).

Daly (1990) lists three rules that we should apply to help define the sustainable limits to material and energy throughput:

- For a *renewable resource* (soil, water, forest, fish, etc.), the sustainable rate of use can be no greater than the rate of regeneration of its source.
- For a *non-renewable* resource (fossil fuel, high grade mineral ore, fossil groundwater, etc.), the sustainable rate of use can be no greater than the rate at which a renewable resource (used sustainably) can be substituted for it.
- For a *pollutant*, the sustainable rate of emission can be no greater than the rate at which that pollutant can be recycled, absorbed, or rendered harmless in its ecosystem 'sink' (where it ends up).

However, as a civilisation we have generally ignored these rules. Indeed, in Australia we have traditionally sought to sell off our mineral resources as quickly as we can. World governments, lobbied by the fossil fuel denial industry, ignore the third rule for carbon dioxide. Accelerating climate change is the result, with huge impacts (IPCC 2014). Conservation of non-renewable resources thus will not happen without rethinking the dominant model of consumer driven economies. It will require a 'new conservation ethic'. The challenge is to reverse incentives, rules and other structures that cause us to be myopic users of resources, and replace them with principles and practices that would make our children (and their children) grateful and proud (Gardner 2013).

So how do we control 'throughput', the growth in use of resources, both non-renewable and renewable? For *non-renewable* resources a depletion quota has been suggested (Daly 1991) or a 'severance tax' at the mine-mouth or well-head (Daly 2008). A depletion quota operates by restricting supply of a resource. The first stage would be a government auction of rights to purchase resources. The total amount of purchase rights to be auctioned in a given year would be determined by legislative decision. A severance tax increases the cost of a resource, encouraging conservation of that resource. Severance taxes can be revenue neutral, by phasing them in while phasing out regressive payroll or sales taxes (Costanza *et al.* 2013). However, while severance taxes in modest form do exist (often we call them 'royalties'), depletion quotas are resisted by governments and are rare. This seems to reflect a fundamental denial of resource scarcity, and the need to limit throughput for both ecological reasons and ethical reasons (regarding intergenerational equity and environmental ethics).

For *renewable resources*, proper holistic pricing of ecosystem services can also reduce overuse (Kumar 2010). However, mostly the ecosystem services benefits are not priced into renewable resources as sold in the free market. In many (probably most) cases these public benefits to society outweigh the private market value of a renewable resource (MEA 2005). As the market mostly doesn't consider the monetary value of ecosystem services vs their current private market price, often renewable resources are degraded and decrease overall well-being. The task of getting society to value and price-in ecosystem services into the free market economy remains an essential (if too slow) work in progress (Kumar 2010).

So we can seek to limit throughput from the production end by depletion quotas and higher severance taxes. However, we can also take action from the consumption end. We often hear of 'dematerialisation' of the economy, and the need for the highest possible *decoupling* of the economy from resource use. This means that we reduce the amount of materials used to make a unit of GDP. How far can we go with this process? Many nations are wasteful in their use of both energy and resources. A key part of dematerialisation is the realisation that we can use much less energy and materials and still have a similar quality of life. This has been variously called Factor 4 (use only 25 per cent of current energy and materials, von Weizsäcker et al. 1998), Factor 5 (use only 20 per cent, von Weizsäcker et al. 2009) and Factor 10 (use only 10 per cent, see Factor 10 Institute website, online, available at: www.factor10-institute.org/). Factor 4 or 5 is certainly possible (if difficult). Business can make a major contribution here by adopting a goal of Factor 5 in their use of materials and energy. However, as discussed in Chapter 4, the idea that we can *totally decouple* our economy from resource use (UNEP 2011) would seem impossible. Another key aspect of reducing material use is to plan products from 'cradle to cradle', so that a product is not just produced to be trashed later. Products must be produced with a view to how they will be reused or recycled into a new product. We should see pollution as a 'symbol of design failure' (Braungart and McDonough 2008). Kopnina and Blewitt (2015, p. 237) summarise the three key principles of 'cradle to cradle' that should inform human design, being: waste equals food; use current solar income (to power a sustainable society), and celebrate Nature's diversity (and learn from it, e.g. biomimicry). Braungart and McDonough (2008) point out we need a process of industrial re-evolution, where we use eco-effectiveness to commit to a new paradigm rather than incrementally improve the old.

The cradle to cradle process and the idea of a 'circular economy' seek to eliminate toxins and non-renewables; eliminate waste; recreate the cycle between urban and agricultural areas, securing mutual nutrient flows; find a dynamic balance between organic and technologically produced products; and identify strategic opportunities for business to engage in the closed loop production on the global scale (Kopnina and Blewitt 2015). To this end we need to prohibit 'planned obsolescence'. This repugnant and wasteful philosophy has gained ground in the last few decades as a 'clever' way of increasing sales. Thus our clothes wear out faster, our tools break, electronic goods don't last as long – all so that we have to buy *more*. Going on with a system that generates massive amounts of waste in the endless spiral of production and consumption only prolongs the faulty system (Braungart and McDonough 2008). In a full world where population is still increasing (as is energy use from fossil fuels), and where our civilisation has an ecological footprint of 1.5 Earths, such a strategy is fundamentally unethical and completely unsustainable.

We need 'extended producer responsibility' (EPR). EPR is a strategy designed to promote the integration of environmental costs associated with goods (throughout their life cycles) into the market price of the products (OECD 2001). It aims to make the manufacturer of the product responsible for the entire life-cycle of the product, and especially for the 'take-back', recycling and final disposal. We also need proper 'life cycle assessment' that covers everything that is significant environmentally (and socially), from the extraction of raw materials, their transformation in the manufacturing processes, to their ultimate disposal, which may include reuse or recycling rather than simply dumping unwanted stuff in landfill sites (Kopnina and Blewitt 2015). Use of the 'Natural Step' approach by business is a valuable approach, as it considers that in a sustainable society, Nature must not be subject to systematically increasing concentrations of substances extracted from the Earth's crust; concentrations of substances produced by society, and degradation by physical means. It makes use of four steps – 'awareness', 'baseline mapping', 'creating a vision, and 'down to action'. Many companies have successfully adopted the Natural Step to improve their sustainability performance and profile (Nattrass and Altomare 2001).

As a society, we also need to put priority on the use and reuse of materials already mined. Braungart and McDonough (2008, p. 52) note we need the '4 Rs' – 'reduce, reuse and recycle and regulate'. But really we need a fifth, 'rethink' (as they point out), we need to rethink as to whether we need to produce the product in that form in the first place (using an eco-effectiveness approach). They point out that recycling is really the aspirin to alleviate our collective hangover of overconsumption. We need to reject throwaway products, and we need to plan for reuse before recycling. More importantly, we need to rethink about whether we want to make a product that pollutes in its production and disposal, and can only be 'downcycled' into less valuable products. So in summary, it is thus possible to dramatically reduce the throughput of materials (and reduce the pollution involved in making things) in our society. However, we need the *political will*.

Beyond 'triple bottom line', 'eco-efficiency' and the 'small and easy' approach

Dematerialisation of our society has the potential to reduce environmental impact, but only if we also *stop endless growth* in population and resource use and our current increasing consumption. If we don't, then any moves to dematerialisation (greater efficiency of resource use) will be swamped by increasing consumption and numbers of consumers. Indeed this has been the case. If we are to demystify sustainability in regard to overconsumption, then we must examine some of the key terms put forward as the 'sustainability solutions' proposed by business. The term 'triple bottom line' was made famous by John Elkington (1998) in 'Cannibals with Forks', who also described it as 'People, Planet, Profit'. These three have often been described as the three pillars of sustainability. The triple bottom line (TBL) is commonly portrayed as *the* key solution to sustainability and certainly underpins business sustainability. TBL is an accounting framework with three dimensions: social, environmental (or ecological) and economic. In this regard it parallels the three strands of sustainability covered here.

The original idea of the '3Ps' in business were the maximisation of 'productivity, profit and power', none of which considered ecological limits (Hill 1992). One should ask if this has been changed by the new '3Ps' of TBL? Opinions differ. Robins (2006) argues that one of the major weaknesses of the TBL is due to the fact that it is an *accounting* framework, so it is based on a monetary-based economic system. Because there is no single way in monetary terms to measure benefits to society and environment (as there is with profit), it does not allow for businesses to easily sum across all three bottom lines. How does one value in dollars the beauty of a wild river or a species threatened with extinction? Similarly, by ranking 'economics' as a separate pillar, it ignores the fact that the economy is a *subset* of society, which is a subset of Nature (see Chapter 4). The final point with TBL is that often it is the 'Trojan Horse' for tokenism. Many organisations have TBL policies and claim their decisions are made using this, when in fact the real decisions are decided first and foremost on economic determinants, with ecological or social considerations an afterthought (Braungart and McDonough 2008, p. 153). Kopnina and Blewitt (2015, p. 189) conclude that sometimes the 3P's (people, profit and planet) simply cannot be balanced, as it is impossible to have your cake and eat it too. Milne and Gray (2013, p. 24) conclude:

> Through the practice of incomplete TBL (aka sustainability) reporting many organizations seem to confuse narrow and incomplete, partial reporting with claims to be reporting on being sustainable, actually being sustainable, or more commonly, with claims to be moving towards sustainability. These claims, we contend, are further exacerbated through the institutional developments [which] may amount to little more than soothing palliatives that, in fact, may be moving us towards greater levels of un-sustainability.

It may thus not be surprising that TBL has been operating for decades, yet we have gone backwards overall regarding sustainability. '*Eco-efficiency*' is the other key term and has been proposed as a key strategy to promote a transformation from unsustainable development to one of 'sustainable development' (Braungart and McDonough 2008, p. 51). It is clearly important in terms of how we dematerialise society. It is based on the concept of creating more goods and services while using fewer resources and creating less waste and pollution. The term was coined by the World Business Council for Sustainable Development (WBCSD) in 'Changing Course' (Schmidheiny 1992) and at the 1992 Earth Summit 'eco-efficiency' was endorsed as a way for companies to implement Agenda 21. Kopnina and Blewitt (2015, p. 185) note that eco-efficiency might not be the long-term solution, for it only works to make the destructive system a bit less so (Braungart and McDonough 2008). In the worst cases, eco-efficiency can make the system that results in overexploitation become more pernicious, because its workings become more subtle and long-term. Braungart and McDonough (2008) advocate instead 'eco-effectiveness', which is an alternative design and production concept, advocating a positive agenda for the conception and production of goods and services by focusing on the development of products and industrial systems that maintain or enhance the quality and productivity of materials through subsequent life cycles.

Eco-efficiency is part of the broad term 'ecological modernisation', which is grounded in the belief in the ability to solve problems by technological advancement. It argues that on the basis of 'enlightened self-interest', supposedly economy and ecology can be favourably combined (Ayres and Simonis 1994). It operates in the belief that environmental problems are caused by poverty and that economic growth, prosperity and equitable distribution of resources are going to solve these problems (Kopnina and Blewitt 2015). Environmental productivity (i.e. productive use of natural resources) is thus seen as a source of future growth and development. Foster (2012) notes that this demands economic reform, with greater efficiency, but involves no break with the dominant structures of capitalist production and consumption or its accumulation imperative. Ecological modernisation includes increases in energy and resource efficiency (eco-efficiency) as well as product and process innovations (e.g. supply chain management, clean technologies). The trap here of course is that without an *acceptance of ecological limits* (now exceeded) and a recognition that population and consumption cannot keep growing, using these tools may not make much difference. As Kopnina and Blewitt (2015, p. 12) conclude: 'the rhetoric of ecological modernisation tends to downplay the essentially insatiable appetites of an increasingly global consumer class'.

One of the consequences of neoliberalism is the belief that market-based solutions will correct environmental problems, focusing political attention on consumer choice and lifestyles (Kopnina and Blewitt 2015). Together, the approaches of TBL, ecological modernisation and eco-efficiency are often portrayed as the 'easy' way to sustainability, and that we can get there in 'small and easy' steps. However, if this was correct, then we would be further along the path to sustainability. Kopnina and Blewitt (2015) point out that if the key problems are overpopulation and overconsumption, then without action on these, eco-efficiency 'results only in tinkering at the margins of the problem without addressing its root causes'. They argue that the 'rebound effect' (or Jevons paradox) suggests that eco-efficiency ultimately leads to more consumption. This paradox is where the consumer response to the introduction of new eco-efficient technologies or products (which save the consumer money) it to actually increase their consumption.

There is a vigorous debate about whether 'greening' our daily individual actions really *does* actually lead people to deeper engagement where they make meaningful changes, or instead lulls them into false security and accomplishment. Are these individual small acts the 'on-ramps' to greater engagement or are they 'dead ends' (Leonard 2013, p. 249)? Foster (2008) points out that people and organisations often operate in 'bad faith' by setting strong targets they secretly know they will not meet. Hill (1992) describes the small and easy approach as 'shallow sustainability', while Rees (2010) notes that most sustainability campaigns emphasise 'simple and painless' solutions. He notes however these are actually 'marginal and ineffective'. Maniates (2013, p. 260) points out that social change does not happen through mass uncoordinated shifts in lifestyles or consumption choices. Hence Maniates notes that 'small and easy' is attractive, plausible – and dead wrong. To deal with consumerism is not just a matter of green marketing, green products and greenwash. Beder devotes a chapter to the 'corporate subversion of the green movement' and of how environmental non-government organisations have been drawn into 'greenwash'. She concludes (Beder 2000, p. 272 and 282):

> But the solutions that such individuals and groups offer will inevitably fit with the existing business/market paradigm, and will provide little challenge to prevailing ideologies or interests.... A new wave of environmentalism is now called for: one that will engage in the task of exposing corporate myths and methods of manipulation.

Rather than the 'small and easy' and greenwash approaches, dealing with consumerism will require:

- strong regulation of corporations;
- banning planned obsolescence;
- enforcing 'cradle to cradle' products;
- enforcing 'take back' laws and EPR;
- clamping down on wasteful packaging;
- constraining rampant advertising;
- applying choice-editing to prohibit the worst of products.

It will be neither small nor easy, but it is necessary, and it is *feasible*. So do we need eco-efficiency? Clearly we do, and this is a key part of moving to dematerialisation and using less energy and producing less greenhouse gases. Do organisations need to consider the environmental and social 'bottom lines' in TBL? Of course they should, as this should inform decision-making in all organisations. Sustainable economic reform and green entrepreneurship, invention, and good design could liberate the creativity of individuals and societies (Kopnina and Blewitt 2015, p. 214). However, are these tools the *key* and primary ways of reaching sustainability? No they are not, certainly not if they are still based on endless growth (which is generally the unwritten subtext). And this is where we must go beyond these convenient buzz words. They are all open to co-option that hides continuing business-as-usual based on endless growth in consumption and resource use. TBL and eco-efficiency are useful tools if used properly, but *only* when they are tools used on the understanding that there are global ecological limits we have exceeded, so that we cannot keep growing as we have been. Sustainability requires that understanding, and without this TBL and eco-efficiency become mere tokenism. Hence why we need to look beyond the terms.

There are, however, promising signs of corporate change. By 2008 some 80 per cent of the world's largest corporations were producing corporate social responsibility (CSR) reports under the Global Reporting Initiative (GRI) (Sukhdev 2013, p. 158). KPMG (2011) studied 378 companies around the world, and reports that the percentage pursuing a sustainability strategy increased from 50 per cent in 2008 to 62 per cent in 2011. The task is now to convert corporate sustainability strategies from tokenism to true reform and action. This is the real task of 'Third Wave' sustainability for corporations, to reinvent themselves, accept the need for a steady state economy, and *aid* sustainability, not hinder it. E. F. Schumacher wrote his seminal book on consumerism 'Small is Beautiful' way back in 1973. It is thus long overdue for us to come to grips with and reject the created ideology of consumerism. The end of the consumer culture will come, willingly or unwillingly, and sooner than we would like to believe. The only question is whether we greet it with a series of alternative ways of orienting our lives and cultures to maintain a good life, even as we consume much less (Assadourian 2013, p. 124). The challenge will be to convince more and more people speedily that further efforts to promote a consumer culture are truly a step in the wrong direction, and that the faster we use our talents and energies to promote a *culture of sustainability*, the better off humanity will be (ibid., p. 124).

Alternatives to the consumer society

Is there an alternative to the consumer society, while still keeping a decent quality of life? In 1960, Cuba was blockaded by the United States (the 'Special Period') and exports dropped by 75 per cent. It had to adapt to severe shortages of oil, medicine and food. However, Cuba now serves as an example of a country that has *thrived* despite having limited fossil fuels. Cuba has low per capita income, yet in quality of life it excels. It is a materially-poor country with 'First World' education, literacy and health care. It has maintained its human services programmes, free education, old age support, basic nutrition and free health care. The WWF Living Planet Report rated Cuba in 2006 as the only country to have genuine sustainable development (Murphy and Morgan 2013). Cuba represents an alternative, where material success (as measured by energy consumption) is secondary, while quality of life is given priority. The message is clear, humanity can survive (and even thrive) in a resource-constrained world if it learns from Cuba's example (Murphy and Morgan 2013). So it's not a case of having to go back and 'live in caves'. We can live a sustainable life with far less consumerism, less 'things', a much smaller ecological footprint – and still have a 'good life'.

Dealing with the heresy of *more*

'More' is not always better, either for people or for stuff. It once used to be, but not in today's world. Yet our current civilisation seems locked in a trance of 'ever-moreism' (Boyden 2004). 'More and more' people is seen as good. More and more things and junk means more consumption. And that is good for the economy isn't it? Actually, in real terms it is not. Beyond a certain point, more people and more consumption brings *uneconomic* growth, as Chapter 4 showed. In real terms it is making society worse off, not better, and decreasing humanity's well-being. It is also decreasing the prospects of future human well-being. We are stealing from posterity (Catton 2012). The big problems of overpopulation and overconsumption 'have no technical fixes but only difficult moral solutions' (Daly 1991, p. 39). Gowdy (2014, p. 40) invites us to do a thought experiment where painlessly humanity's population is returned to a few hundred million and ecosystems restored. He notes however that if we keep the ideology of growth, accumulation and expansion, then we would be back to the current system in a few decades, too many people consuming too much and the Earth's life support systems teetering on collapse. Endless growth is the true heresy of 'more' that we must confront and change.

Campbell (2012, p. 52) concludes we can break the spiral of silence about overpopulation by showing that it (1) it makes it impossible for the poor to escape poverty and ecological degradation, (2) high fertility is not due to women's desire to have more children, (3) fertility can decline when women are given freedom to control their fertility via family planning. Dealing with overpopulation will be hard, given the baggage that goes with it. Dealing with consumerism will be even harder as it is actually easier to change family size than patterns of consumption (Campbell 2012, p. 45). This may be partly because the advertising industry spends $500 billion a year urging us to consume more, where we are continually told we need to (Assadourian 2013, p. 115). As the growth economy falters (as it is), we will increasingly be told that the 'only solution' is to consume more and more junk. It is a false solution, and the sooner we realise this, the better. It took a decade or two to embed consumerism into the US psyche, which then got exported around the world. It will take time to move away from rampant consumerism, to move back to 'thriftiness', to a more equitable level of consumption. Most (but not all) corporations will oppose a change, as will the advertising industry, and probably also the car industry and the fast food industry. The list of those in opposition will undoubtedly stretch on and on. Sadly, 'time' is one of the things we don't have 'more' of, but the least of. We must act right now to cure our consumer addiction.

Most governments will sit on their hands and ignore the problems of overpopulation and overconsumption, while some will actively promote population growth and consumption in an attempt to increase the GDP and avoid a recession. So, if we are going to break the addiction, it will have to *come from us*. We will have to demand an ecologically sustainable population, demand a ban on 'planned obsolescence', demand we (as a society) consume less, demand our governments reduce waste, and demand that corporate law is reformed. It won't be easy, but then the consequences of ignoring overpopulation and overconsumption will be far worse. These are the key drivers of *un*sustainability. Some argue for action on one but not the other, but the two are conjoined-twins in terms of environmental impact. We cannot accept the problems of one and ignore the other. We will not reach true sustainability if we cannot recognise the problems of both, move past denial, and solve these drivers of unsustainability. 'Sustainability' (if it is to mean more than tokenism) has to mean dealing with overpopulation and overconsumption. Demystifying sustainability means facing up to the two key drivers of unsustainability, the two key barriers holding us back – overpopulation and overconsumption. Without tackling these, any purported 'sustainability strategies' will remain pure tokenism.

References

Assadourian, E. (2010) 'The rise and fall of consumer cultures', in *2010 State of the World: Transforming Cultures from Consumerism to Sustainability*, eds. L. Starke and L. Mastny, London: Earthscan.

Assadourian, E. (2013) 'Re-engineering cultures to create a sustainable civilization', in *State of the World 2013: Is Sustainability Still Possible?*, ed. L. Starke, Washington: Island Press.

Ayres, R. U. and Simonis, U. E. (1994) *Industrial Metabolism: Restructuring for Sustainable Development*, Tokyo: UN University Press.

BBC (2006) 'Earth is too crowded for Utopia', *BBC News online*, 6 January 2006, online, available at: http://news.bbc.co.uk/2/hi/science/nature/4584572.stm.

BCSD (1993) 'Getting Eco-efficient', report of the Business Council for Sustainable Development first Antwerp Eco-Efficiency Workshop, Geneva, online, available at: http://infohouse.p2ric.org/ref/23/22632.pdf (accessed 2 December 2014).

Beder, S. (2000) *Global Spin: The Corporate Assault on Environmentalism*, revised edition, Melbourne: Scribe.

Boyden, S. (2004) *The Biology of Civilisation: Understanding Human Culture as a Force in Nature*, Sydney: UNSW Press.

Braungart, M. and McDonough, W. (2008) *Cradle to Cradle: Remaking the Way we Make Things*, London: Vintage Books.

Brown, L. (2011) *World on the Edge: How to Prevent Environmental and Economic Collapse*, New York: W. W. Norton and Co.

Brown, L. (2012) *Full Planet, Empty Plates: the New Geopolitics of Food Scarcity*, New York: Norton.

Butler, T. (2012) 'Colossus versus liberty: a bloated humanity's assault on freedom', in *Life on the Brink: Environmentalists Confront Overpopulation*, eds. P. Cafaro and E. Crist, Georgia, US: University of Georgia Press, pp. 160–171.

Campbell, M. (2012) 'Why the silence on population?', in *Life on the Brink: Environmentalists Confront Overpopulation*, eds. P. Cafaro and E. Crist, Georgia, US: University of Georgia Press, pp. 41–55.

Catton, W. (1982) *Overshoot: the Ecological Basis of Revolutionary Change*, Chicago: University of Illinois Press.

Catton, W. (2012) 'Destructive momentum: could an enlightened environmental movement overcome it?', in *Life on the Brink: Environmentalists Confront Overpopulation*, eds. P. Cafaro and E. Crist, Georgia, US: University of Georgia Press, pp. 16–28.

Collins, P. (2010) *Judgment Day: the Struggle for Life on Earth*, Sydney: UNSW Press.

Costanza, R., Alperovitz, G., Daly, H., Farley, J., Franco, C., Jackson, T., Kubiszewski, I., Schor, J. and Victor, P. (2013) 'Building a sustainable and desirable economy-in-society-in-nature', in *State of the World 2013: Is Sustainability Still Possible?*, ed. L. Starke, Washington: Island Press.

Crist, E. (2012) 'Abundant Earth and the population question', *Life on the Brink: Environmentalists Confront Overpopulation*, eds. P. Cafaro and E. Crist, Georgia: University of Georgia Press, pp. 141–151.

Daly, H. (1990) 'Toward some operational principles of sustainable development', *Ecological Economics*, vol. 2, pp. 1–6.

Daly, H. (1991) *Steady State Economics*, Washington: Island Press.

Daly, H. (2008) 'A steady-state economy: a failed growth economy and a steady-state economy are not the same thing; they are the very different alternatives we face', 'thinkpiece' for the Sustainable Development Commission, UK, 24 April, online, available at: http://steadystaterevolution.org/files/pdf/Daly_UK_Paper.pdf (accessed 2 December 2014).

Dietz, R. and O'Neill, D (2013) *Enough is Enough: Building a Sustainable Economy in a World of Finite Resources*, San Francisco: Berrett-Koehler Publishers.

Ehrlich, P. (1968) *The Population Bomb*, New York: Ballentine Books.

Ehrlich, P. (2013) Personal communication from Prof. Paul Ehrlich at the 2013 Fenner Conference on the Environment 'Population, Resources and Climate Change', Canberra Australia, online, available at: http://population.org.au/fenner-conference-2013-summing.

Ehrlich, P., Ehrlich, A. and Holdren, J. (1977) *Ecoscience: Population, Resources, Environment*, New York: W. H. Freeman and Co.

Ehrlich, P. R. and Ehrlich, A. H. (2014) 'It's the numbers, stupid!', in *Sustainable Futures: Linking Population, Resources and the Environment*, eds J. Goldie and K. Betts, Melbourne: CSIRO Publishing.

Ekins, P. (1991) 'The sustainable consumer society: a contradiction in terms?', *International Environmental Affairs*, vol. 3, pp. 243–257.

Elkington, J. (1998) *Cannibals with Forks: the Triple Bottom Line of 21st Century Business*, Canada: New Society Publishers.

Ellul, J. (1975) *The New Demons*, New York: Seabury Press.

Engelman, R. (2012a) 'Trusting women to end population growth', in *Life on the Brink: Environmentalists Confront Overpopulation*, eds. P. Cafaro and E. Crist, Georgia, US: University of Georgia Press.

Engelman, R. (2012b) 'Nine population strategies to stop short of 9 billion', in *State of the World 2012: Moving Toward Sustainable Prosperity*, ed. L. Starke, Washington: Island Press.

Engelman, R. (2013) 'Beyond sustainababble', in *State of the World 2013: Is Sustainability Still Possible?*, ed. L. Starke, Washington: Island Press.

Erb, K., Haberl, H., Krausmann, F., Lauk, C., Plutzar, C., Steinberger, J., Muller, C., Bondeau, A., Waha, K. and Pollack, G. (2009) 'Eating the planet: feeding and fueling the world sustainably, fairly and humanely – a scoping study', Social ecology working paper no. 116, Vienna: Institute of Social Ecology and Potsdam Institute for Climate Impact Research.

FAO (2011) *The State of the World's Land and Water Resources for Food and Agriculture (SOLAW)*, Rome: Food and Agriculture Organisation.

Flavin, C. (2010) 'Preface' to *State of the World 2010: Transforming Cultures from Consumerism to Sustainability*, eds. L. Starke and L. Mastny, New York: Worldwatch Institute/Earthscan.

Foster, J. (2008) *The Sustainability Mirage: Illusion and Reality in the Coming War on Climate Change*, London: Earthscan.

Foster, J. B. (2012) 'The Planetary Rift and the New Human Exemptionalism: A Political-Economic Critique of Ecological Modernization Theory', *Organization and Environment*. vol. 25, no. 3, pp. 211–237.

Gardner, G. (2013) 'Conserving non-renewable resources', in *State of the World 2013: Is Sustainability Still Possible?*, ed. L. Starke, Washington: Island Press.

GFN (2013) 'World footprint: do we fit on the planet?', Global Footprint Network, online, available at: www.footprintnetwork.org/en/index.php/GFN/page/world_footprint/ (accessed 20 September 2013).

Gowdy, J. (2014) 'Governance, sustainability, and evolution', in *State of the World 2014: Governing for Sustainability*, ed. L. Mastny, Washington: Island Press.

Graff, J. (2010) 'Reducing work time as a path to sustainability', in *State of the World 2010: Transforming Cultures from Consumerism to Sustainability*, eds. L. Starke and L. Mastny, New York: Worldwatch Institute/Earthscan.

Hamilton, C. (2010) *Requiem for a Species: Why We Resist the Truth About Climate Change*, Australia: Allen and Unwin.

Hardin, G. (1968) 'The tragedy of the commons', *Science*, vol. 162, no. 3859, pp. 1243–1248.

Hill, S. B. (1992) 'Ethics, sustainability and healing', Talk to Alberta Round Table on the Environment and Economy and the Alberta Environmental Network, 11 June, online, available at: www.beliefinstitute.com/article/ethics-sustainability-and-healing (accessed 23 June 2014).

Hulme, M. (2009) *Why We Disagree About Climate Change: Understanding Controversy, Inaction and Opportunity*, Cambridge: Cambridge University Press.

IPCC (2014) Summary for Policymakers, Fifth Assessment Report, International Panel on Climate Change, online, available at: http://ipcc-wg2.gov/AR5/images/uploads/WG2AR5_SPM_FINAL.pdf (accessed 16 June 2014).

Kolankiewicz, L. (2012) 'Overpopulation versus biodiversity: How a plethora of people produces a paucity of wildlife', in *Life on the Brink: Environmentalists Confront Overpopulation*, eds. P. Cafaro and E. Crist, Georgia, US: University of Georgia Press.

Kolankiewicz, L. and Beck, R. (2001) 'Forsaking fundamental: the environmental establishment abandons US population stabilisation', Washington: US Centre for Immigration Studies, online, available at: www.cis.org/articles/2001/forsaking/toc.html (accessed 3 July 2014).

Kopnina, H. and Blewitt, J. (2014) *Sustainable Business: Key Issues*, London: Routledge.

KPMG (2011) *Corporate Sustainability: A Progress Report*, online, available at: www.kpmg.com/US/en/IssuesAndInsights/ArticlesPublications/Documents/iarcs-eiu-corporate-sustainability.pdf.

Kumar, P. (2010) *The Economics of Ecosystems and Biodiversity: Ecological and Economic Foundations*, London: Earthscan.

Layard, R. (2005) 'The national income: a sorry tale', in *Growth Triumphant: the 21st Century in Historical Perspective*, ed. R. Easterlin, Michigan: Ann Arbor.

Lebow, V. (1955) 'Price competition in 1955', *Journal of Retailing*, Spring, 1955, online, available at: http://classroom.sdmesa.edu/pjacoby/journal-of-retailing.pdf (accessed 9 March 2012).

Leonard, A. (2013) 'Moving from individual change to societal change', in *State of the World 2013: Is Sustainability Still Possible?*, ed. L. Starke, Washington: Island Press.

Maniates, M. (2013) 'Teaching for turbulence', in *State of the World 2013: Is Sustainability Still Possible?*, ed. L. Starke, Washington: Island Press.

MEA (2005) *Living Beyond Our Means: Natural Assets and Human Wellbeing, Statement from the Board, Millennium Ecosystem Assessment*, United Nations Environment Programme, online, available at: www.millenniumassessment.org.

Meadows, D., Randers, J. and Meadows, D. (2004) *Limits to Growth: the 30-year Update*, Vermont: Chelsea Green.

Meyerson, F. (2007), 'Rising carbon emissions call for a population policy', *Bulletin of Atomic Scientists*, 3 December 2007, online, available at: www.thebulletin.org/population-and-climate-change/rising-carbon-emissions-call-population-policy (accessed 2 December 2014).

Milne, M. and Gray, R. (2013) 'W(h)ither ecology? the triple bottom line, the global reporting initiative, and corporate sustainability reporting', *Journal of Business Ethics*, vol. 118, no. 1, pp. 13–29.

Monbiot, G. (2009) 'The population myth', online, available at: www.monbiot.com/archives/2009/09/29/the-population-myth/ (accessed 22 July 2014).

Moore, J. and Rees, W. (2013) 'Getting to one-planet living', in *State of the World 2013: Is Sustainability Still Possible?*, ed. L. Starke, Washington: Island Press.

Murphy, P. and Morgan, F. (2013) 'Cuba: lessons from a forced decline', in *State of the World 2013: Is Sustainability Still Possible?*, ed. L. Starke, Washington: Island Press.

Nattrass, B. and Altomare, M. (2001) *The Natural Step for Business*, Gabriola Island, Canada: New Society Publishers.

O'Connor, M. and Lines, W. (2008) *Overloading Australia: How Governments and Media Dither and Deny on Population*, Sydney: Envirobook.

OECD (2001) 'Extended producer responsibility: a guidance manual for governments', Organisation for Economic Cooperation and Development, online, available at: www.oecd-ilibrary.org/environment/extended-producer-responsibility_9789264189867-en.

Ostrom, E. (1990) *Governing the Commons: the Evolution of Institutions for Collective Action*, Cambridge: Cambridge University Press.

PM (2010) 'Capacity population', Population Matters leaflet, online, available at: http://populationmatters.org/documents/capacity_leaflet.pdf (accessed 15 November 2011).

Prugh, T. (2013) 'Getting to true sustainability', in *State of the World 2013: Is Sustainability Still Possible?*, ed. L. Starke, Washington: Island Press.

Rees, W. (2008) 'Toward sustainability with justice: are human nature and history on side?', in *Sustaining Life on Earth: Environmental and Human Health through Global Governance*, ed. C. Soskolne, New York: Lexington Books.

Rees, W. (2010) 'What's blocking sustainability? Human nature, cognition and denial, *Sustainability: Science, Practice and Policy*, vol. 6, no. 2 (ejournal), online, available at: http://sspp.proquest.com/archives/vol. 6iss2/1001–012.rees.html (accessed 21 June 2014).

Robins, F. (2006) 'The challenge of TBL: a responsibility to whom?', *Business and Society Review*, vol. 111, no. 1, pp. 1–14.

Ryerson, W. (2012) 'How do we solve the population problem', in *Life on the Brink: Environmentalists Confront Overpopulation*, eds. P. Cafaro and E. Crist, Georgia, US: University of Georgia Press, pp. 240–254.

Schmidheiny. S. (1992) *Changing Course*, Cambridge, MA: MIT Press.

Schumacher, E. (1973) *Small is Beautiful*, New York: Harper Torchbooks.

Staples, W. and Cafaro, P. (2012) 'For a species right to exist', in *Life on the Brink: Environmentalists Confront Overpopulation*, eds. P. Cafaro and E. Crist, Georgia, US: University of Georgia Press, pp. 283–300.

Starke, L. and Mastny, L. (2010) *State of the World 2010: Transforming Cultures from Consumerism to Sustainability*, New York: Worldwatch Institute/Earthscan.

Sukhdev, P. (2010) Preface to *The Economics of Ecosystems and Biodiversity: Ecological and Economic Foundations*, ed. P. Kumar, London: Earthscan.

Sukhdev, P. (2013) 'Transforming the corporation into a driver of sustainability', in *State of the World 2013: Is Sustainability Still Possible?*, ed. L. Starke, Washington: Island Press.

Tacey, D. (2000) *Re-enchantment: the New Australian Spirituality*, Australia: Harper Collins.

UN (2012) *World Population Prospects; The 2012 Revision*, online, available at: http://esa.un.org/unpd/wpp/Documentation/pdf/WPP2012_highlights.pdf (accessed 2 December 2014).

UNEP (2011) *Towards a Green Economy: Pathways to Sustainable Development and Poverty Eradication*, United Nations Environment Programme, online, available at: www.unep.org/greeneconomy.

Victor, P. (2008) *Managing Without Growth: Slower by Design, not Disaster*, Cheltenham, UK: Edward Elgar.

Washington, H. (1991) *Ecosolutions: Solving Environmental Problems for the World and Australia*, Tea Gardens, NSW, Australia: Boobook Publications.

Washington, H. (2013) *Human Dependence on Nature: How to Help Solve the Environmental Crisis*, London: Earthscan (Routledge).

Washington, H. and Cook, J. (2011) *Climate Change Denial: Heads in the Sand*, London: Earthscan.

Weeden, D. and Palomba, C. (2012) 'A post-Cairo paradigm: both numbers and women matter', in *Life on the Brink: Environmentalists Confront Overpopulation*, eds P. Cafaro and E. Crist, Georgia, US: University of Georgia Press.

Weizsäcker, E. von, Hargroves, K., Smith, M., Desha, C. and Stasinopoulos, P. (2009) *Factor 5: Transforming the Global Economy through 80% Increase in Resource Productivity*, London: Earthscan.

Weizsäcker, E. von, Lovins, A. and Lovins, H. (1998) *Factor 4: Doubling Wealth, Halving Resource Use*, London: Earthscan.

Westra, R. (2008) 'Market society and ecological integrity: theory and practice', in *Sustaining Life on Earth: Environmental and Human Health through Global Governance*, ed. C. Soskolne, New York: Lexington Books.

Wijkman, A. and Rockstrom, J. (2012) *Bankrupting Nature: Denying our Planetary Boundaries*, London: Routledge.

Wilkinson, R. and Pickett, K. (2010) *The Spirit Level: Why Equality is Better for Everyone*, London: Penguin Books.

Zencey, E. (2013) 'Energy as a master resource', in *State of the World 2013: Is Sustainability Still Possible?*, ed. L. Starke, Washington: Island Press.

8
WORLDVIEW AND ETHICS IN 'SUSTAINABILITY'

> *A human being is a part of the whole called by us universe, a part limited in time and space. He experiences himself, his thoughts and feeling as something separated from the rest, a kind of optical delusion of his consciousness. This delusion is a kind of prison for us, restricting us to our personal desires and to affection for a few persons nearest to us. Our task must be to free ourselves from this prison by widening our circle of compassion to embrace all living creatures and the whole of nature in its beauty. Nobody is able to achieve this completely but the striving for such achievement is in itself a part of the liberation and a foundation for inner security.*
>
> (Albert Einstein 1950)

We cannot demystify 'sustainability' in any meaningful way unless we talk about worldview, ethics and values (and even spirituality). Despite this, many books about sustainability do not mention the first two, and only touch obliquely on 'values'. Perhaps nothing illustrates the 'tokenistic' side of 'sustainability' more than this failure. Some sustainability classics such 'Cradle to Cradle' (Braungart and McDonough 2008, p. 11) even argue that 'we won't solve problems if they are seen as ethical'. However, I side with environmental ethicists who argue that we won't solve environmental problems unless we talk about ethics (e.g. Rolston 2012). Worldview and ethics are central, for if society does not question and address these, then it simply won't change and hence will not reach sustainability. In previous chapters, we found our economy was broken, our society was broken, and that the Nature that supports us is *breaking*. There is a reason for this, because the anthropocentric 'mastery of Nature' worldview of Western society is unsustainable. We touched on this in Chapter 1, and found our ancestor's worldview was very different. We enlarged on this in Chapter 4 in regard to 'growthism'. We found that the 'mastery of Nature' view was impoverishing the natural world (and twisting ecological theory) in Chapter 5. We found in Chapter 6 that our society itself was a casualty, that we don't live in a happy society, but one that is anxious, competitive and consumerist. Perhaps we should consider the underlying cause? We discuss here worldview and ethics, anthropocentrism vs ecocentrism, the intrinsic value of Nature, and ideologies. There is what could be called a 'great divide' between those who are anthropocentric and ecocentric. Sadly, this divide has probably never been as great as it is today. Humanity is now more divorced from the natural world than ever before (Louv 2005, p. 2). How did we get to this stage? Is our current Western worldview a cultural maladaptation (Boyden 2004)?

Worldview, ethics, values and ideologies

In academia you will repeatedly come across the terms 'worldview', 'paradigms' and 'ideologies'. Worldview is self-explanatory, it is how we view the world. A paradigm may be seen as related but somewhat narrower than a worldview, it is generally described as a 'distinct concept or thought pattern'. Cavagnaro and Curiel (2012, p. 165) argue that paradigms are 'like spectacles through which people perceive a situation'. The lenses are formed by a person's principles, knowledge and experiences, and by the value and norms of the surrounding society. An *ideology* is a set of conscious (and unconscious) ideas that constitute one's goals, expectations, and actions. An ideology is a comprehensive vision, a way of looking at things. A related term is a 'meme', which is an idea, behaviour, or style that spreads from person to person within a culture. Harich (2012) argues that infection by 'false memes' (e.g. 'climate change is not real') is one of the key problems we face in creating change for sustainability.

You can see that it is easy to switch from worldview to paradigm to ideology in the same discussion, and this is often the case. However, arguably worldview is the wider view, followed by paradigm, with ideology the most specific (which is how I treat them here). I speak here of various 'ideologies'. One of my reviewers for this book noted: 'but the author has his own ideology'. And indeed we all have worldviews, ethics and values. I certainly come from an ecocentric worldview and should make that clear (though by now this should be obvious!). One could call this an ideology, though for me it is broader and hence a worldview. I find an ecocentric worldview scientifically, ethically and spiritually responsible. Finally it should of course be noted that 'ethics' is about what is 'right' or 'wrong' conduct. This involves what we value. Why is worldview or paradigm so important? Donnella Meadows (1997) explains this beautifully:

> The shared idea in the minds of society, the great unstated assumptions . . . constitute that society's deepest set of beliefs about how the world works. . . . Growth is good. Nature is a stock of resources to be converted to human purposes. . . . Those are just a few of the paradigmatic assumptions of our culture, all of which utterly dumfound people of other cultures.

Cavagnaro and Curiel (2012, p. 168) similarly explain why worldview is central:

> Because everything else follows from the way we look at reality, the moment we are able to embrace a new, sustainable, world view our minds will open to new possibilities; we will be able to understand which other steps are needed and find ways to actually take them.

As Meadows (1997) concluded:

> People who manage to intervene in systems at the level of a paradigm hit a leverage point that totally transforms systems.... In a single individual it can happen in a millisecond. All it takes is a click in the mind, a new way of seeing.

So what is the problem with our current modernist 'mastery of Nature' worldview? Berry (1988) argues:

> We can break the mountains apart; we can drain the rivers and flood the valleys. We can turn the most luxuriant forests into throwaway paper products.... We can pollute the air with acids, the rivers with sewage, the seas with oil – all this in a kind of intoxication with our power for devastation.... And why? To increase the volume and speed with which we move natural resources through the consumer economy to the junk pile or the waste heap. Our managerial skills are measured by our ability to accelerate this process.... If the environment is made inhospitable for a multitude of living species, then so be it. We are supposedly, creating a technological wonderworld.... But our supposed progress toward an ever-improving human situation is bringing us to a wasteworld instead of a wonderworld.

Rees (2008) notes that:

> The world presently has the wealth, human capital and natural resources to execute a smooth transition to global sustainability out of mutual self interest, yet we do not act.... Despite forty years of organised environmentalism, two world summits on environment and development, repeated warnings by scientists and the emergence of 'sustainable development' as a mainstream mantra, global society continues its drive toward ecological disaster and geopolitical chaos.

He further notes that humanity got to this point through unconscious expansionist tendencies, reinforced by the social construction of both a 'perpetual growth myth' and increasingly a 'global consumer myth'. Those who would guide humanity to sustainability are therefore pitted against formidable biological and cultural imperatives. Greer (2009) concludes that civilisation is sleepwalking toward the abyss, as we fail to grasp the most basic elements of ecological reality. The relentless pursuit of economic efficiency and material wealth lacks any moral or ethical compass, so we 'continue to erode natural systems that demand consideration of time horizons far beyond shareholder reports and election cycles' (Soskolne *et al.* 2008). Daly and Cobb (1994, p. 21) note in regard to ethics that 'Before this generation is the way of life and the way of death'. They conclude that society is 'living by an ideology of death and accordingly we are destroying our own humanity and killing the planet'. Accordingly, part of demystifying sustainability is to realise that Band-Aid solutions will not work. We need to change society's worldview. We need an 'ecological worldview' (Catton 1982), also called an 'ecosophy', a philosophy of ecological harmony (Næss 1989), a 'biounderstanding' worldview (Boyden 2004), a 'Great Turning' (Macy 2012) or 'compassionate retreat' (Brown and Schmidt 2014). Our current human-centred worldview has been called 'Manifest Destiny' (McCright and Dunlap 2000) and 'Defiance', where we defy and deny evidence of the environmental crisis (McKibben 2006). It sees Nature as just a group of resources that only have value for human use. But why is a new worldview so essential? Because without it we will not remove the Cornucopian cataracts from our eyes, and we will remain blind to the underlying real causes of our failure to reach sustainability (as noted long ago by Catton 1982).

The alternative is to adopt an *ecocentric* approach. To value Nature for itself and see the natural world as something 'sacred', of which we are a part. We should ask if we are ego-centric or ecocentric? Is the Universe just about *us* and our consumption, or is it about sharing our planet with the wondrous evolved diversity of life? Over the past 200 years, the Western modernism we discussed in Chapter 1 has impoverished the natural world we share, and brought us to the brink of tremendous further loss, as we found in Chapter 5. This is the 'elephant in the room' we can no longer afford to ignore. We must now be 'realists'. Some 30 years ago, Catton (1982) argued there are five ways to approach 'ecologically inexorable change'. These five ways were based on whether people accepted the 'circumstances' of environmental impact, but also the 'consequences' of the need to change. These were *realism* (we accept the circumstances and consequences); *cargoism* (accept circumstances but disregard consequences); *cosmeticism* (disregard circumstances but partially accept consequences); *cynicism* (disregard both circumstances and consequences); and *ostrichism* (circumstances and consequences are both denied). It is time to be a 'realist' and accept both the circumstances of the predicament we have created and the consequences of our lack of action. That means thinking about how we view the world, and whether the modernist Western worldview is totally unsustainable.

Ethics are related to values, and Zerubavel (2006) speaks about the silence of denial about ethics:

> 'The best way to disrupt moral behaviour' notes political theorist C. Fred Alford 'is not to discuss it and not to discuss not discussing it'. 'Don't talk about ethical issues' he facetiously proposes 'and don't talk about our not

talking about ethical issues'. As moral beings we cannot keep non-discussing 'undiscussables'. Breaking this insidious cycle of denial calls for an open discussion of the very phenomenon of undiscussability.

People *are* starting to reconsider their worldview, and have been revolting against the narrow focus of modernism for decades. The ethical dimensions of climate change are starting to be recognised. Anglican Bishop David Atkinson (2008) argues:

> Climate change is ... opening up for us ... questions about human life and destiny, about our relationship to the planet and to each other, about altruism and selfishness, about the place of a technological mindset in our attitude to the world, about our values, hope and goals, and about our obligations for the present and for the future. These are moral and spiritual questions.

Recently the Catholic Jesuit Social Justice and Ecology Secretariat produced a report, 'Healing a Broken World' (Alvarez 2011). It noted:

> The deterioration of the environment as a result of human activity has taken on a decisive importance for the future of our planet and for the living conditions of coming generations. We are witnessing a growing moral consciousness regarding this reality.... Nevertheless, we are still in need of a change of heart. We need to confront our inner resistances and cast a grateful look on creation, letting our heart be touched by its wounded reality and making a strong personal and communal commitment to healing it.

Some 72 nations now mention environmental rights in their constitutions (Engel 2008), the most famous being Bolivia and Ecuador. There is also the Pachamama movement (see Pachamama Alliance website, online, available at: www.pachamama.org/) that seeks to use traditional wisdom to reach sustainability. So there is a growing recognition that ethical change is *needed*. Collins (2010) believes discussion now has to move to the 'moral' sphere, that we face a massive, overarching moral problem, bigger than war, more serious than financial meltdowns. He believes we have to talk in a language that shows what we are doing to Nature is 'sinful', that we are committing ecocide. We now need to adopt a fundamental moral principle that the good of the planet must come 'before everything else' (Collins 2010). As Berry (1994) has noted 'the ecological imperative is not derivative from human ethics' but the other way around. There are many practical and useful things that can be done without changing our worldview, especially a transition to renewable energy (Diesendorf 2014), but if we don't also change our worldview, we are likely to get ourselves re-entangled in the same mess as today. Fundamentally, we need to tackle the 'isms' of modernism, industrialism and consumerism.

Can we change our worldview? Some people feel threatened by this question. Of course, it has recently actually been changing for the worse, with the spread of neoliberalism, based on belief in unfettered markets (Martin 2013, p. 269). However, things are happening, and a clean industrial revolution is on the way (McNeil 2009; Diesendorf 2014). Similarly, people's views of how humans relate to Nature are changing. Many people do 'care' about the Earth. This is where the conservation and environment movements came from, this is where the desire to buy 'green' products comes from. This is where the drive to sustainability comes from. This is why the Global Alliance for the Rights of Nature was formed (see website, online, available at: http://therightsofnature.org/). This is why there is a legal move towards 'Earth jurisprudence' (Cullinan 2003, 2014).

Anthropocentrism

If I was to run around shouting 'Me! Me! Me!', you would soon get sick of me. Yet that is essentially what humanity has been doing over recent history. Humanity has become self-obsessed, we focus on ourselves, or at least the majority of us in Western society do. We focus on our society, our economy, and only lastly on the Nature that supports us. Even Green political parties fall into this trap. Anthropocentrism regards humanity as the central element in the Universe, *ecocentrism* is instead focused on a Nature-centred system of values, and accepts that humanity is part of Nature and must treat it with responsibility and respect.

One of the main issues addressed by environmental ethics is the dilemma of anthropocentrism vs ecocentrism, and of intrinsic value of Nature vs utilitarian (i.e. just for human use) or instrumental value (value as a means to acquiring something else) (Norton 1995; Zack 2002; Rolston 2012). Anthropocentrism has dominated modern societies since at least the sixteenth century (Smith 1998). The universality and insidious aspect of anthropocentrism in modern culture has been attested to by Næss (1973), Godfrey-Smith (1979), Smith (1998) and Rolston (2012). Anthropocentrism of course gets disseminated around the world by globalisation (what Braungart and McDonough 2008, p. 119 call 'a tide of sameness'). This serves only to distance economic life further from agriculture and Nature than did early capitalism (Westra 2004). There has also been a clear tendency for philosophers to focus primarily on the human mind. Næss (1973) noted that some scholars argue that most forms of human knowledge are 'inherently anthropocentric', that we are incapable of acknowledging ecosystem importance. It has been said that by being human 'we can only be anthropocentric: we seek our own good, not what we suppose is nature's' (Lowenthal 1964). However, this is actually a sterile ideological claim, not a fact. Taylor (1986) points out that humans *can* take an animal's standpoint 'without a trace of anthropocentrism', and make judgements of what is desirable from that standpoint.

A central assumption of modernist Western moral thought is that value can be ascribed to the non-human world only in so far as it is good 'for the sake of humans' (Godfrey-Smith 1979). The Western modernist attitude toward Nature thus has a decidedly anthropocentric bias. However, an important distinction in this debate is known as the 'Anthropocentric Fallacy'. This explains that just because humans can only perceive Nature by 'human' senses does not mean we cannot 'attribute' intrinsic value to it (Fox 1990; Eckersley 1992). By way of comparison, men (even white men) are quite capable of cultivating a non-sexist or non-racist consciousness. They don't 'have' to be sexist or racist, and can clearly attribute value to women and dark-skinned people. Similarly, humans are quite capable of cultivating a non-anthropocentric consciousness (Fox 1990), and attributing intrinsic value to Nature. To understand the environment will always involve human senses and imagination, but Smith (1998) asks does that mean 'humans should always be the measure of all things?'. Just because we are human does not in fact mean we *have* to be egotists and focus on ourselves. As Chapter 1 showed, humanity has not always been this way. If we are going to reach sustainability, we cannot afford to stay that way.

Western culture has become ingrained with a 'doctrine of inherent human superiority', and this has become 'an unfounded dogma of our culture' (Taylor 1986). Taylor argues for a 'biocentric outlook', but notes that this cannot be 'proven' as such, since worldviews are not deductive systems or theories. It is not so much a question of 'proof', as of taking an ethical stance and worldview. If one takes an ecocentric worldview, then it can readily be justified, but first one has to take that view. The worldview one takes determines one's core underlying ethics and values. It has been noted that if we conceive of Nature as a 'machine', then that allows the human mind to retain a god-like position 'outside' of the world. If the worldview of 'Nature as a machine' rose to prominence in the seventeenth century (following the work of Descartes and Newton) due to its compatibility with a 'divine creator', 'it remains in prominence today largely due to the deification of human powers that it promotes' (Abram 1992). Seeing Nature as a machine, something less than ourselves, thus allows us to pretend we are something *more*. Crist (2012) notes there is a particularly virulent strain of anthropocentrism that is 'human supremacy', the belief that humans are the superior life form of the planet and Earth's entitled 'owners'. Human supremacy has become so entrenched that the wondrous diversity of life is reduced down to just 'resources' or 'natural capital'. Seeing the beauty of the world as just 'things' (resources for our use) has become seen as 'normal'. Crist (2012) argues that the concept of resources has come to be 'a gaping wound on the face of language' and engraved the delusion of human supremacy into 'common sense', science and politics. She believes that if we continue the delusion of human supremacy it will extinguish the possibility of 'yet-to-be-imagined (sane, harmonious, beautiful) ways of being on Earth'.

Hay (2002) notes that there is currently something of a backlash within society *against* ecocentrism and in favour of anthropocentrism, with greater emphasis on social justice and emergent democracy. This backlash demonstrates the insidious nature of anthropocentrism as an assumption in Western society, now sadly being disseminated around the world by globalisation. Humanism and humanitarianism have been linked to anthropocentrism, arguing that it affirms the human side of the Nature/culture pair, and that humanism must come to terms with the denied non-human side (Plumwood 2001). Humanism has arguably helped us to lose touch with ourselves as beings that are *also natural*, and have their roots in the Earth. Hence Plumwood (2001) notes that anthropocentric culture often portrays Nature as 'passive or dead'.

Anthropocentrism in 'sustainable development'

Helen Kopnina (2012) traces the differences between the older idea of 'environmental education' and the newer education for sustainable development. Leopold (1949) argued that humans should act to protect Nature for the good of other species. Environmental education thus came from a conservation perspective, and the Belgrade Charter (UNEP and UNESCO 1975) made clear that environmental education should 'favour a return to harmony with nature ... and to give emphasis to the relationship of belonging, replacing the anthropocentric worldview by an ontocentric worldview' (ontocentric means to be centred on 'being' or reality). However, education for sustainable development places stronger emphasis on human rights. Spring (2004) points out the problem of the anthropocentric approach advocated by mainstream sustainable development. Kopnina (2012) notes that environmental protection in sustainable development is 'seen as an afterthought to all other pressing human issues such as equality, fair distribution of natural resources and human rights'. She concludes that the mainstream academic discussion around sustainable development maintains an instrumental and anthropocentric worldview.

Ärlemalm-Hagsér and Sandberg (2011) in a study of pre-school teachers' comprehension of sustainable development found the teachers thought sustainable development focused on human rights, democracy, gender equality, morals and ethics. Environment came as an afterthought. Kopnina concludes that sustainable development now masks an *inherent anthropocentric bias*. She believes that the current encouragement of plural interpretations of sustainable development may allow corporate and political elites to exclude ecocentric perspectives from consideration. This pluralism in sustainable development represents the voice of a single species (humans), and marginalises the rest of life. Kopnina (2012) notes that deep ecology is only mentioned in passing in sustainable development academic discussion, suggesting that the majority of sustainable development scholars are of a lighter shade of green. She concludes that the continued prioritisation of an anthropocentric perspective within sustainable development can lead to the denial of the older aim of environmental education, which was to solve key problems such as the biodiversity extinction crisis. Anthropocentrism in sustainable development thus means that people and profit are being put before planet. Kopnina (2012) concludes 'academic relativism about education for sustainable development might in fact be undermining the efforts of educating citizens in the importance of valuing and protecting the environment'.

Ideologies – modernism

Chief amongst the things that aid anthropocentrism are our ideologies, and chief amongst these is the 'modernism' we discussed in Chapter 1. Modernism took a strong anthropocentric view of the world as being essentially a resource for human use (resourcism). Crist (2012) summarises an ecocentric response to the idea of seeing Nature just as a resource:

> What is deeply repugnant about such a civilization is not its potential for self-annihilation, but its totalitarian conversion of the natural world into a domain of resources to serve a human supremacist way of life, and the consequent destruction of all the intrinsic wealth of its natural places, beings, and elements. 'Project Human Takeover' has proceeded acre by acre, island by island, region by region, and continent by continent, reaching its current global apogee with the final loss of wild places and the corollary sixth mass extinction underway.

Intrinsic value and the revolt against modernism

However, modernism and the dominance of a utilitarian view of Nature did not completely triumph. The idea that Nature has 'intrinsic value', a right to exist for itself, irrespective of its use to humanity, was resurrected by the writings of the Romantic poets such as Wordsworth and Coleridge. Later it was espoused by the visionaries Thoreau, Muir and Leopold (Oelschlaeger 1991), as discussed in Chapter 1. When discussing 'intrinsic value' one must be careful as to what one *means*. The term carries a lot of philosophical baggage, and the debate resembles the religious debate about 'how many angels can fit on the head of a pin?'. It is complicated by a past history of attack by pragmatism, debates about objectivism and subjectivism, and by claims equating it to hedonism (Monist 1992). Much of the discussion is about the term 'intrinsic' itself rather than whether Nature as such has value in itself. Several philosophers agonise about whether certain parts of Nature have *more* intrinsic value than others (Monist 1992). Taylor (1986) argues that intrinsic value is the wrong term, that we should use 'inherent worth'. However, to skirt this can of worms, I use 'intrinsic value' here for the simple idea that the non-human world has value (worth) irrespective of whether it is of use to humanity. Accordingly, we should respect it.

The intrinsic values and 'Rights of Nature' have been recognised in the 'World Conservation Strategy' (IUCN 1980) and in the Earth Charter. The 'Millennium Ecosystem Assessment' (MEA 2005) and the UNEP project 'The Economics of Ecosystems and Biodiversity' (Kumar 2010) both acknowledge it (even if only in passing). Regarding intrinsic value, Rolston (1985) argues 'such values are difficult to bring into decisions; nevertheless, it does not follow that they ought to be ignored'. Taylor (1986) argues we should *respect* Nature even if we don't 'love' unattractive parts of it. Hargrove (1992) argues that non-living Nature such as geodiversity should also be seen as having intrinsic value. Discussion of intrinsic value is mostly relegated to specialty journals, particularly ethics journals, and thus does not adequately penetrate the whole academic (or indeed popular) literature.

Zimmerman (2001) discusses intrinsic value in an analytical philosophical approach, though almost nothing is said about the intrinsic value of Nature itself. The philosophical journal the *Monist* (1992) produced a special edition on 'The Intrinsic Value of Nature'. The real problem is that Nature has been mapped philosophically as a moral blank space, as 'value-free' in and of itself (Rolston 2012). However, if we accept the intrinsic value of Nature, this makes it that much easier to change our worldview to one that respects Nature and allows us to live sustainably within it.

Deep ecology

Deep ecology has been one academic stream that has championed intrinsic value. Philosopher Arne Næss (e.g. Næss 1973, 1984, 1989, 2008) and Bill Devall and George Sessions (1985) were leaders of this environmental philosophy characterised by its advocacy of the inherent worth of living beings, regardless of their use to human needs, and advocacy for a radical restructuring of modern society. Deep ecology developed a platform of eight principles (Devall and Sessions 1985):

1   The well-being and flourishing of human and non-human life on Earth have value in themselves (synonyms: intrinsic value, inherent value). These values are independent of the usefulness of the non-human world for human purposes.
2   Richness and diversity of life forms contribute to the realisation of these values and are also values in themselves.
3   Humans have no right to reduce this richness and diversity except to satisfy vital human needs.
4   The flourishing of human life and cultures is compatible with a substantial decrease of the human population. The flourishing of non-human life requires such a decrease.
5   Present human interference with the non-human world is excessive, and the situation is rapidly worsening.
6   Policies must therefore be changed. These policies affect basic economic, technological, and ideological structures. The resulting state of affairs will be deeply different from the present.
7   The ideological change is mainly that of appreciating life quality (dwelling in situations of inherent value) rather than adhering to an increasingly higher standard of living. There will be a profound awareness of the difference between big and great.
8   Those who subscribe to the foregoing points have an obligation directly or indirectly to try to implement the necessary changes.

Deep ecology is thus a philosophical worldview that accepts the intrinsic value of Nature, supports wilderness protection, understands that human population must decrease, that the economy and technology must change – and that this is urgent. However, the dominance of anthropocentrism within both society and academia has meant that deep ecology has sadly struggled to increase its influence. A move back towards deep ecology within both academia and society would assist the transformation to a meaningful 'sustainability'.

Ideologies – postmodernism

So under modernism, Nature had no value of its own, its only value was as something to be *used* by humanity. Postmodernism has been called a 'geography of ideas' that developed in opposition to modernism. It appears resistant to being defined (Butler 2002). Instead of espousing clarity and certitude, postmodernism commits itself to ambiguity and relativity (Crotty 1998). The postmodernist penchant to accept that everything is 'relative', and that we 'co-create reality', has meant that the work of some in academia can be used by deniers to create their own distorted view of environmental science. It has increased scepticism about the existence of objective truth. Cohen (2001) notes that this 'epistemic relativism turns scientific facts into mere "social constructions"'. Some themes within postmodernism relevant to sustainability are:

1 'reason' as defined by Western society is itself suspect (Derrida 1966);
2 the denial of grand narratives (theories organising overall meaning) (Lyotard 1992);
3 that it is impossible to prove the *real* from a 'simulacra' (Baudrillard 1983);
4 concern for the 'other' (Kristeva 1992).

One can legitimately ask whether many of the postmodernist streams of thought have been any less anthropocentric than modernism. For many postmodernists, wild Nature as an independent entity seems to be reduced to just a discourse operating 'within human minds' (Baudrillard 1983). Gare (1995) observed that postmodernism has demonstrated the many problems of modernism, while being 'powerless to oppose them'. Postmodernism is thus good at analysing the problems of modernism, but poor at finding solutions to them. As an ideology, it (like its predecessor modernism), is thus not helping us to reach sustainability. The postmodernist definition of 'the other' may seem hopeful, but all too often it seems to be limited to the human species, rather than extending compassion to the rest of the 'more-than-human' world (Abram 1996). It is time to expand our moral circle *beyond* our own species (Singer 1981).

The 'suspicion of reason' clearly poses problems for a rational assessment of the environmental crisis. The 'denial of all grand narratives' means that postmodernism does not supply a 'story' to replace modernism with an over-arching sustainability ethic. Just as evolution has been portrayed as a grand narrative (Docherty 1992), so too the environmental crisis can be seen as 'just another' grand narrative. Hence postmodernism can deny the possibility of constructing any coherent meta-narrative of the environmental crisis (Collins 2010). Cohen (2001) notes that some postmodernist statements, such as that morality and values are relativistic, are 'simply ludicrous'. He argues that while they stayed in academia, such statements were 'harmless fun', but when they circulate in mass culture they aid denial. Deniers can then claim they are 'simply offering an alternative version of history' (Cohen 2001). It seems to me that this leads to more fiddling as Rome burns, that it doesn't help us reach sustainability. We need something more. We do in fact need a grand narrative, an all-encompassing vision of sustainability, of repairing the Earth (Berry 1999).

Kopnina (2012) notes that post-structuralism (part of postmodernism) argues that education should have 'no meaning and no ends'. The relativism and pluralism in postmodernism (especially deconstructive postmodernism) can lead to the dissolution of moral responsibility for non-human species and lead to scepticism and nihilism (Callicott 1999). Kopnina concludes that the pluralistic, liberal or emancipatory approach of education for sustainable development signals a 'scholarly departure from "real-world" dilemmas concerned with environmental degradation'. The postmodern approach within 'sustainable development' is thus largely still anthropocentric and tends to ignore ecocentric values and shy away from finding real solutions to real problems. Indeed, its focus on relativism tends to mean postmodernism will argue that all solutions are equally valid, rather than focus on the key areas needed. Hence society (and much of academia) continues to fiddle as Rome burns.

Questioning reality and 'Nature scepticism'

The questioning of 'the real' serves merely to distance humanity even more from Nature. Nature is real, our ecological dependency on Nature is real, and human actions really *are* degrading the ecosystems that support real human societies. Postmodernists often speak of 'co-creating reality', when actually there is a 'real' reality (one very worthy of respect), which humans interpret differently. Some postmodernist academics argue that 'it is now difficult to sustain a position of "naïve realism". In scholarly circles it is difficult to suggest that the world exists outside our construction of it' (Reason and Torbert 2001). Cleary under this description, I am a 'naïve realist'! However, if Nature is not 'real' to us, then we are unlikely to feel a responsibility towards it, accord it ethical status, or think much about ecological sustainability. The postmodernist questioning of reality thus continues the anthropocentric view of the world developed by modernism. This has led to a significant debate within academia about the land being a 'human artefact'. A key problem here is the distinction between *influencing* the land (as all indigenous peoples did) and *creating* it (which is anthropocentric as it places all the emphasis on 'human' creation). For example, it has been asserted that 'virgin forests and wilderness areas are in part artefacts of previous burns, both natural and anthropogenic ... tropical forests are "both artefact and habitat"' (Gomez-Pampa and Kaus 1992).

Similarly, in Australia it has been stated: 'here is no wilderness, but there are cultural landscapes ... those of Aboriginal people, present and past, whose relationships with the environment shaped even the reproductive mechanisms of forests' (Langton 1996). So is Nature just a human artefact? Conservation biologist Michael Soulé (1995) notes the claim that humans 'invented the forest' ignores that species' geographical distributions are determined largely by ecological tolerances, geological history and climate. Similarly, wilderness is not simply a cultural construct 'devised to mirror our own broken nature', but is a home to all that is wild, a blank space on the map 'that illustrates human restraint' (Tempest Williams 1999). Bryant (1995) notes that:

> If we turn our regard for nature more and more into clever philosophical word games, if we begin to think that we are intellectually creating nature rather than physically participating in it, we are in danger of losing sight of the real wolves being shot by real bullets from real aeroplanes, of real trees being clearcut, of real streams being polluted by real factories.

Rolston (2001) argues that wilderness:

> created itself long before civilisation ... wildness a state of mind? Wildness is what there was before there were states of mind ... it seems that the main idea in nature is that the natural is not a human construct. Intentional, ideological construction is exactly what natural entities do not have: if they had it, they would be artefacts. The main idea in nature is that nature is not our idea.

Hay (2002) adds an essential question:

> Why should it be assumed that the smallest incursion of culture into nature constitutes the end of nature? It is just as logical to argue the opposite – that because trees grow in London's parks ... London has ceased to be part of the realm of culture, and has become nature. The fact is that there are natural processes and there are cultural processes, and in any place the mix is likely to be uneven.

Given the predominance of anthropocentrism within academia, this 'breaking down' of the human/Nature split seems to have been attempted through what Plumwood (2003) has called 'Nature scepticism', by denying that 'Nature' as such exists. Some argue that because humans influence Nature, it thus somehow 'becomes human'. Others argue the need to equate Nature with culture, where again Nature (the non-human) disappears within culture (the human). Thus one is left with only 'cultural landscapes' and no natural or wild areas. Wilderness, as a place where Nature comes first, has fallen foul of such ideology (see my website, online, available at: www.wildernesstruths.com), being dismissed as a 'romantic legacy', or an attempt to maintain the Nature/culture dualism. The response broadly shared by Gare (1995), Rolston (2001) and Plumwood (2001) seems useful. Humans and their culture *are* a part of Nature, but we are a 'distinctive' part. We thus need a conception of Nature that allows humans to be essentially cultural beings, while still seeing them as part of (and within) Nature (Gare 1995). By recognising the 'other' of wildness, we bring culture and Nature together (Rolston 2001). We can thus continue to use words such as 'culture' and 'Nature', just as we can recognise that any landscape will be a result of a spectrum of natural *and* cultural influences (Hay 2002).

There is also something happening beyond Nature scepticism, which one could call 'biophobia' or fear of life, or at least a distaste for the messy living world. Part of this is the idea in Tennyson's poem that Nature is: 'red in tooth and claw'. This claim that Nature is harsh, competitive and brutal ignores that Nature is also full of *cooperation*, that our very cells are a testament to the cooperation of two sorts of organisms who united (the mitochondria inside our cells contain bacterial DNA). Rolston (1992) encapsulates this:

> Nature is random, contingent, blind, disastrous, wasteful, indifferent, selfish, cruel, clumsy, ugly, struggling, full of suffering, and, ultimately, death? This sees only the shadows, and there has to be light to cast shadows. Nature is orderly, prolific, efficient, selecting for adapted fit, exuberant, complex, diverse, renews life in the midst of death, struggling through to something higher.... We miss the panoramic creativity when we restrict value to human consciousness.

A sense of *wonder*

How then can we improve our worldview and ethics? Most of us (at least at times) feel a *sense of wonder* at Nature. Many of us can remember this wonder from childhood. A yearning for a 'sense of place' is a perennial human longing. All peoples need a sense of 'my country', of belonging to a sustaining landscape they respond to in care and love (Rolston 2012). Wilderness can be valued as a place that restores one's 'sense of wonder' in life (Washington 2002). This can also be called a 're-enchantment' of the land (Tacey 2000). Part of this sense of wonder is a feeling of being 'one' with the land, of *belonging* (Thomashow 1996). This issue fascinated me to the extent that I self-published the book 'A Sense of Wonder' (Washington 2002). So, what is this sense of wonder at the natural world? Is it something innate we are born with, or can anybody learn it? Is it something more common in different races, or merely different societies? Does the fact that only 20 per cent or so of people (in Western society) get interested in environmental issues mean that only that percentage can *feel* the sense of wonder? Alternatively, does the growing interest in the environment in young people mean that the number who can feel a sense of wonder is growing? These are important questions to ponder.

It is common for environmentalists to list all sorts of 'scientific' arguments about 'why' a natural area should be saved. It must be saved because it is a reservoir of biodiversity. It should be saved because there are 'X' threatened species. It should be saved because it has unique geodiversity, because it provides ecosystem services. It should be saved because it provides clean air and water to society. I know this list well, as I have run campaigns to protect wild places using similar arguments myself. These arguments are of course valid, and this book lists similar reasons for valuing Nature. However, it seems that through our history of scientific analysis we have lost the ability to see Nature as an 'interlocking whole' (Collins 2010). Infrequently do we argue that a natural area should be saved because it is a place of 'wonder' we love, or because it is 'sacred'. One's words could then be dismissed as 'emotional'. It is strange that we have come to such a pass, if we cannot get emotional about the land we love – what *can* we get emotional about? Or is our passion dead?

The sense of wonder I am talking about here is a connection with the land. As such it must be considered 'spiritual', even if one does not call it religious. An agnostic, a Christian, a Buddhist, a Hindu, a Muslim ... all can feel a sense of wonder, even if they call it by varying names (Washington 2002). What I am talking about here is not something that should be trivialised and put on the shelf with the books on fairies. I am talking about the fundamental relationship of humanity with the land that has nurtured us for millions of years. I am talking about one of the deepest and abiding loves of them all: the love of the land. There are also high points in our sense of wonder. I have called them 'transcendent moments' (Washington 2002) while others have called them a 'hierophany' or epiphany (Oelschlaeger 1991). Such moments can change one's life. Wilderness advocate John Muir was escaping from the military draft in 1864, and was near Lake Huron in Canada. He wandered into a dark swamp. There he came upon a cluster of rare white orchids. They were so beautiful that he later wrote (Corcoran 2001):

> I never before saw a plant so full of life; so perfectly spiritual. It seemed pure enough for the throne of its Creator. I felt as if I were in the presence of superior beings who loved me and beckoned me to come. I sat down beside them and wept for joy.

The sense of wonder is also about 'empathy'. One must let down one's guard, open oneself up and let all one's senses absorb the beauty of the natural world. Aboriginal Elder Miriam-Rose Ungunmerr (see Tacey 2000) explains that there is a word in the Ngangikurungkurr language 'dadirri', which is 'something like what you (white people) call "contemplation"'. Tacey (2000) believes this is a spirituality of deep seeing and deep listening. This would seem to me to be part of the empathy that allows us to feel the sense of wonder. Ungunmerr believes that the gift of Dadirri is 'perhaps the greatest gift we can give to our fellow Australians' (Tacey 2000). The writings of other scholars, and my own lived experience, force me to conclude that a sense of wonder is an important part of the human psyche. It could play a key role in helping us reach true sustainability. It can help to solve 'Nature Deficit Disorder' (Louv 2005). Reaching sustainability will require that we use our emotional intelligence as much as our intellectual one (Hempel 2014). Our sense of wonder at Nature is thus not just 'one more value', but something of central importance, something that gives us *meaning*.

Anti-spirituality in Western culture

Part of the problem is that modern Western society does not just ignore environmental ethics and a sense of wonder, it actively opposes a deep spiritual connection to Nature. Tacey (2000) argues:

> The consumerist society reinforces our shrunken empty status. It is vitally important for capitalism that we continue to experience ourselves as empty and small, since this provides us with the desire to expand and grow, and this desire is what consumerism is based on. Consumerism assumes that we are empty but permanently unable to fulfil our spiritual urge to expand. It steps into the vacuum and offers its own version of expansion and belonging.... If we stopped believing in the myth of our shrunken identity, the monster of consumerism would die, because it would no longer be nourished by our unrealized spiritual urges. Therefore true spirituality ... is extremely subversive of the status quo. This is why consumer society is keen to debunk or ridicule true spirituality.

This is very relevant to the discussion of consumerism in Chapter 7. Why do so few people today seem to feel a sense of wonder? Why is it relegated to children, and only survives in a minority of adults (at least in Western society)? Where did we go so wrong that our society acts to *suppress* the sense of wonder at Nature, rather than celebrate and foster it? It is not like losing your handkerchief in the park, or misplacing your keys. It is the fundamental alienation of oneself from the natural world. Clearly something is very wrong for us to have come to this state. Clearly our whole worldview must be at fault (Berry 1988), must be out of kilter. How could we let our sense of wonder, that connection that gives life meaning, slip away? It is clear that Nature in Western society is no longer sacred, no longer the 'Earth Mother' as it was in the 'old sustainability' (see Chapter 1). It has become just a 'resource' for human use. The underlying worldview of Western society is thus at great pains to suppress a sense of wonder and a spiritual connection with Nature. This is something that desperately needs to change if we are to reach any meaningful sustainability.

How do we bridge the great divide?

So we need to bridge the divide between humanity and Nature, and acknowledge its intrinsic value. We are a part of Nature, but a special part that also has culture, and the two can and must mesh in an ecologically sustainable way. We must challenge the anthropocentric assumption, and develop an 'ethic of ecological obligation', a 'land ethic' that widens the moral community to include the land, as

Leopold (1949) taught. It has been argued a first step in recognising an 'enlarged' moral community is the evolution of 'empathy' (Godfrey-Smith 1979). To change our culture, Taylor (1986) thought we needed an inner change in our moral beliefs, from anthropocentrism to biocentrism and respect for Nature. The recognition of the rights of other species and ecosystems to exist for themselves can be seen as an acceptance of 'humility' (Noss 1991), and indeed as a gesture of 'planetary modesty' (Nash 2001).

As a society, we need to talk about intrinsic value and acknowledge our anthropocentrism. Our education system generally doesn't discuss environmental ethics, nor in general do our universities. They need to. Hence why the situation is worrying that Kopnina (2012) reports, where education for 'sustainable development' has become more anthropocentric than 'environmental education' used to be (see Chapter 3). If education for sustainable development remains anthropocentric, the chances of us solving the environmental crisis are poor. One way to break down anthropocentrism is to rekindle our 'sense of wonder'. Berry (1988) suggests that there is hope in breaking free from anthropocentrism:

> Yet the beginning of an intimacy can be observed. The very intensity of our inquiry into the structure and functioning of the natural world reveals an entrancement with this natural world.... We are constantly drawn toward a reverence for the mystery and the magic of the Earth and the larger Universe with a power that is leading us away from our anthropocentrism to this larger context as our norm of reality and value.

The fact that humans can mourn the extinction of another species (as many people do) shows that we do *care*. As Aldo Leopold (in Meadows 1999) noted: 'For one species to mourn the death of another is a new thing under the sun'. Rolston (2012, p. 26) in his illuminating 'A New Environmental Ethics' argues that environmental ethics is now vital for the survival of life on Earth. We are searching for an urgent world vision, for an 'Earth ethics'. It is better to build our cultures in intelligent harmony with the way the world *is*, rather than try to 'take control' as Western society has been doing. Rolston (2012, p. 46) asks if we want Nature to end? He wonders:

> What kind of planet do we want? What kind of planet can we get? Maybe we also ought to ask: What kind of planet do we have? What kind of planet ought we to want? Maybe we ought to develop our capacities for gratitude, wonder, respect and restraint. Maybe we live on a wonderland Earth that we ought to celebrate as much as to develop.

He believes we have to choose either the nature of our 'apartnesss' or the nature of our 'partness'. After all, we are 'Earthlings', our integrity is inseparable from Earth integrity. People and their Earth have *entwined destinies* (Rolston 2012, p. 39). Assadourian (2013, p. 296) notes that environmentalists have 'stolen fear, guilt and sin from religion', but they have left behind celebration, hope and redemption. The problem is that fear without hope, guilt without celebration, and sin without redemption is a model that fails to inspire or motivate. Environmentalists must create a more comprehensive philosophy, complete with an ethics, cosmology, even stories of redemption, that could deeply affect people and change the way they live.

The Earth is not something we outgrow or rebuild and 'manage to our liking', it is the ground of our being. We humans too belong on this planet, it is our home (Rolston 2012, p. 222). Environmental ethics invites awakening to the greater story of which humans are a part. This is what 'human dominion' in the Bible should have meant (Rolston 2012, p. 59). This takes humans past resource use into 'residence'. Being a 'resident' is something more than maximum exploitation. It takes us past 'management' to ethics. We need to be liberated from our egoism, from humanism, into a transcending overview that sees the Earth as a blessed land, exuberant with life, filled with beauty and storied history (Rolston, 2012, p. 59). Environmental ethics calls for 'seeing' non-humans, for seeing the life that cannot say 'I', but in which there is value. Environmental ethics advances beyond human ethics in that it can treat as ends others besides humans. Our planet of promise is now a planet in peril. Rolston (2012, p. 222) concludes that 'At this rupture of history, environmental ethics is vital for today and tomorrow'.

Cavagnaro and Curiel (2012) in their book 'The Three Levels of Sustainability' have an interesting focus in the last third of their book, which is on 'Leadership for sustainability'. They discuss the ideas of 'spiritual intelligence' (p. 256) and 'spiritual capital', as proposed by Zohar and Marshall (2004). Cavagnaro and Curiel (2012, p. 259) argue that 'leaders for sustainability' are: 'those men and women who widen their circle of compassion to embrace all living creatures and the whole of nature on their leadership journey for a more sustainable world'. They conclude that it is through the development of 'our capacity to reach a "Care for all" perspective by developing a spiritual intelligence that sustainability can be achieved at all levels'. Similarly, Kopnina and Blewitt (2015) in 'Sustainable Business' have a strong focus on environmental ethics and intrinsic value. Given that books on sustainability rarely discuss worldview, ethics or spirituality, it is refreshing (and timely) to see other books on sustainability are now placing worldview and ethics (and even 'spiritual intelligence') as a central part of reaching sustainability.

One key way to bridge the 'great divide' is to rejuvenate your sense of wonder. Ways of doing this are (Washington 2002):

- Be there with Nature! Belong in the land.
- Take your children and friends to wild places so they can see the natural world as it really is, and bond with it.
- Take time to ponder ... whether this is called meditation or empathy or prayer or contemplation or 'dadirri' or just sitting somewhere 'at one with the world'.
- Keep your imagination, creativity and artistic expression alive. In these you find the well-spring of your 'being' which renews your sense of wonder.

- Cherish the imagination of your children and youth in general, especially so they can survive the turmoil of puberty. Young people *need* our spiritual help while they undertake their own spiritual journeys.
- Encourage your empathy on a sunny day. Find a beautiful spot and let your defences down and empathise with the natural world. Meditate or just watch and ponder. Perhaps you too will find, as Thoreau (1854) did, that: 'Every little pine needle expanded and swelled with sympathy and something kindred to me'.

Surely this is not too much, not so hard? The potential drawbacks of doing this are negligible. The potential gains are enormous – a world where you can live and let live, a world where you are in harmony, a world where you *belong*.

So to reach sustainability, society today is faced with a complex problem of 'conscious evolution' (Ehrlich and Ornstein 2010) or cultural evolution to create a thoughtful, ecologically-literate human, motivated by a sense of responsibility to the planet (Soskolne *et al.* 2008). The change in worldview is unlikely to come from governments and business (though there are exceptions such as Ray Anderson (1998) of Interface). It certainly won't come from a media owned by conservative corporations. It is sadly also unlikely to come from the current education system (Ehrlich and Ornstein 2010). That leaves *us* – 'we the people'. This is not as hopeless as some might think. Ecologist Paul Ehrlich notes there are thresholds in human behaviour when cultural evolution moves rapidly: 'When the time is ripe, society can be transformed virtually overnight' (in Daily and Ellison 2002). When enough of us change our worldview, then governments, business and the education system will follow. To quote Gandhi: 'We must become the change we want to see in the world'. It is past time for our species to break free from its self-absorption. It is time to do what Berry (1988) suggests and be 'drawn toward a reverence for the mystery and the magic of the Earth'. Humans are not the measure of all things. It is time to reconnect to Nature. It is time to return to our past wonder at (and respect for) Nature. Time to become fully human in the widest, most connected and holistic sense. Time to come *home*.

References

Abram, D. (1992) 'The mechanical and the organic: on the impact of metaphor in science', *Wild Earth*, vol. 2, no. 2, pp. 70–75.

Abram, D. (1996) *The Spell of the Sensuous*, New York: Vintage Books (Random House).

Alvarez, P. (2011) 'Healing a broken world' (special report of the taskforce on ecology), Jesuit Social Justice and Ecology Secretariat, Rome, online, available at: www.sjweb.info/sjs/PJnew (accessed 2 December 2014).

Anderson, R. (1998) *Mid-course Correction: Towards a Sustainable Enterprise: the Interface Model*, Atlanta, GA: Pererinzilla Press.

Ärlemalm-Hagsér, E., and Sandberg, A. (2011) 'Sustainable development in early childhood education: in-service students comprehension of the concept', *Environmental Education Research*, vol. 17, no. 2, pp. 187–200.

Assadourian, E. (2013) 'Re-engineering cultures to create a sustainable civilization', in *State of the World 2013: Is Sustainability Still Possible?*, ed. L. Starke, Washington: Island Press.

Atkinson, D. (2008) *Renewing the Face of the Earth: a Theological and Pastoral Response to Climate Change*, London: Canterbury Press.

Baudrillard, J. (1983) 'Simulacra and simulations', in *Jean Baudrillard, Selected Writings*, ed. M. Poster (1988), Stanford: Stanford University Press.

Berry, T. (1988) *The Dream of the Earth*, San Francisco: Sierra Club Books.

Berry, T. (1994) quoted by McDonough, S. (1994) *Passion for the Earth: Christian Vocation to Promote Justice, Peace and the Integrity of Creation*, London: Geoffrey Chapman.

Berry, T. (1999) *The Great Work: Our Way into the Future*, New York: Bell Tower.

Boyden, S. (2004) *The Biology of Civilisation: Understanding Human Culture as a Force in Nature*, Sydney: UNSW Press.

Braungart, M. and McDonough, W. (2008) *Cradle to Cradle: Remaking the Way we Make Things*, London: Vintage Books.

Brown, L. and Schmidt, J. (2014) 'Living in the Anthropocene: business as usual, or compassionate retreat', in Mastny, L. (ed.) *State of the World 2014: Governing for Sustainability*, Washington: Island Press.

Bryant, P. (1995) 'Constructing nature again (Australian Society of Literature on the Environment Network)', in *Main Currents in Western Environmental Thought*, ed. P. Hay (2002), Sydney: UNSW Press, pp. 24–25.

Butler, C. (2002) *Postmodernism: a Very Short Introduction*, Oxford and New York: Oxford University Press.

Callicott, J. (1999) 'Moral monism in environmental ethics defended', in *Beyond the Land Ethic: More Essays in Environmental Philosophy*, Albany, NY: State University of New York Press.

Catton, W. (1982) *Overshoot: the Ecological Basis of Revolutionary Change*, Chicago: University of Illinois Press.

Cavagnaro, E. and Curiel, G. (2012) *The Three Levels of Sustainability*, Sheffield: Greenleaf Publishing.

Cohen, S. (2001) *States of Denial: Knowing About Atrocities and Suffering*, New York: Polity Press.

Collins, P. (2010) *Judgment Day: the Struggle for Life on Earth*, Sydney: UNSW Press.

Corcoran, P. (2001) 'John Muir', in *Fifty Key Thinkers on the Environment*, ed. J. Palmer, London: Routledge, pp. 131–136.

Crist, E. (2012) 'Abundant Earth and the population question', in *Life on the Brink: Environmentalists Confront Overpopulation*, eds. P. Cafaro and E. Crist, Georgia: University of Georgia Press, pp. 141–151.

Crotty, M. (1998) *Foundations of Social Research: Meaning and Perspective in Research Process*, Sydney: Allen and Unwin.

Cullinan, C. (2003) *Wild Law: a Manifesto for Earth Justice*, Totnes, Devon: Green Books.

Cullinan, C. (2014) 'Governing people as members of the Earth community', in *State of the World 2014: Governing for Sustainability*, ed. L. Mastny, Washington: Island Press.

Daily, G. and Ellison, K. (2002) *The New Economy of Nature: the Quest to make Conservation Profitable*, Washington: Island Press.

Daly, H. and Cobb, J. (1994) *For the Common Good: Redirecting the Economy Toward Community, the Environment, and a Sustainable Future*, Boston: Beacon Press.

Derrida, J. (1966) 'Structure, sign and play in the discourse of the human sciences', in *Critical Theory Since Plato*, ed. H. Adams, New York: Harcourt Brace Jovanovich.

Devall, B. and Sessions, G. (1985) *Deep Ecology: Living as if Nature Mattered*, Utah: Gibbs M. Smith.

Diesendorf, M. (2014) *Sustainable Energy Solutions for Climate Change*, London: Routledge.
Docherty, T. (1992) *Postmodernism: a Reader*, New York: Harvester Wheatsheaf.
Eckersley, R. (1992) *Environmentalism and Political Theory: Toward an Ecocentric Approach*, London: UCL Press.
Ehrlich, P. and Ornstein, R. (2010) *Humanity on a Tightrope: Thoughts on Empathy, Family and Big Changes for a Viable Future*, New York: Rowman and Littlefield.
Einstein, A. (1950) Quote from a letter of condolence to Norman Salit on March 4, 1950, as shown in the Albert Einstein Archives, 61–226, Hebrew University of Jerusalem, online, available at: http://aefind.huji.ac.il/vufind1/Record/EAR000025410/TOC#tabnav (accessed 2 December 2014).
Engel, J. (2008) 'A covenant of covenants: a federal vision of Global 27 governance for the twenty-first century', in *Sustaining Life on Earth: Environmental and Human Health through Global Governance*, ed. C. Soskolne, New York: Lexington Books.
Fox, W. (1990) *Toward a Transpersonal Ecology: Developing New Foundations for Environmentalism*, Boston: Shambhala.
Gare, A. (1995) *Postmodernism and the Environmental Crisis*, London and New York: Routledge.
Godfrey-Smith, W. (1979) 'The value of wilderness', *Environmental Ethics*, vol. 1, pp. 309–319.
Gomez-Pampa, A. and Kaus, A. (1992) 'Taming the wilderness myth', *Bioscience* vol. 42, no. 4, pp. 271–279.
Greer, J. M. (2009) *The Ecotechnic Future: Envisioning a Post-peak World*, Canada: New Society Publishers.
Hargrove, E. (1992) 'Weak anthropocentric intrinsic value', *Monist*, vol. 75, no. 2, pp. 208–226.
Harich, J. (2012) 'The dueling loops of the political powerplace', Twink organization paper, 5 May, online, available at: www.thwink.org/sustain/articles/005/DuelingLoops_Paper.htm (accessed 2 December 2014).
Hay, P. (2002) *Main Currents in Western Environmental Thought*, Sydney: UNSW Press.
Hempel, M. (2014) 'Ecoliteracy: Knowledge is not enough', in *State of the World 2014: Governing for Sustainability*, ed. L. Mastny, Washington: Island Press.
IUCN (1980) *World Conservation Strategy: Living Resource Conservation for Sustainable Development*, Switzerland: International Union for Conservation of Nature and Natural Resources.
Kopnina, H. (2012) 'Education for sustainable development (ESD): The turn away from "environment" in environmental education?', *Environmental Education Research*, vol. 18, no. 5, pp. 699–717.
Kopnina, H. and Blewitt, J. (2015) *Sustainable Business: Key Issues*, London: Routledge.
Kristeva, K. (1992) 'The other of language', in *The Judgement of Paris: Recent French Theory in a Local Context*, ed. K. Murray, Sydney: Allen and Unwin, pp. 23–37.
Kumar, P. (2010) *The Economics of Ecosystems and Biodiversity: Ecological and Economic Foundations*, London: Earthscan.
Langton, M. (1996) 'The European construction of wilderness', *Wilderness News* (Australia), vol. 143, pp. 16–17.
Leopold, A. (1970 [1949]) *A Sand Country Almanac, with Essays on Conservation from Round River*, New York: Random House.
Louv, R. (2005) *Last Child in the Woods: Saving our Children from Nature-deficit Disorder*, London: Atlantic Books.
Lowenthal, D. (1964) 'Is wilderness "paradise enow"?: images of nature in America', *Columbian University Forum*, vol. 7, pp. 34–40.
Lyotard, J. R. (1992) 'Answering the question: what is postmodernism?', in *Postmodernism: A Reader*, ed. T. Docherty, New York: Harvester Wheatsheaf.

McCright, A. and Dunlap, R. (2000) 'Challenging global warming as a social problem: an analysis of the conservative movement's counter-claims', *Social Problems*, vol. 47, no. 4, pp. 499–522.

McKibben, B. (2006) *The End of Nature*, New York: Random House.

McNeil, B. (2009) *The Clean Industrial Revolution: Growing Australian Prosperity in the Greenhouse Age*, Sydney: Allen and Unwin.

Macy, J. (2012) 'A wild love for the world', Joanna Macy interview by Krista Tippett, 'On Being', American Public Media, 1 November, online, available at: www.onbeing.org/program/wild-love-world/61 (accessed 11 September 2013).

Martin, B. (2013) 'Effective crisis governance', in *State of the World 2014: Governing for Sustainability*, ed. L. Mastny, Washington: Island Press.

MEA (2005) *Living Beyond Our Means: Natural Assets and Human Wellbeing, Statement from the Board, Millennium Ecosystem Assessment*, United Nations Environment Programme, online, available at: www.millenniumassessment.org.

Meadows, D. (1997) 'Places to intervene in a system: in increasing order of effectiveness', *Whole Earth*, Winter 1997, pp. 78–84.

Meadows, D. (1999) 'Sand Country almanac fifty years later', online, available at: www.donellameadows.org/archives/sand-county-almanac-fifty-years-later/ (accessed 26 July 2014).

Monist (1992) 'The Intrinsic Value of Nature, Special Edition', *The Monist*, vol. 75, no. 2.

Næss, A. (1973) 'The shallow and the deep, long-range ecology movement: a summary', *Inquiry*, vol. 16, pp. 95–100.

Næss, A. (1984) 'A defence of the deep ecology movement', *Environmental Ethics*, vol. 6, p. 266.

Næss, A. (1989) *Ecology, Community and Lifestyle*, Cambridge: Cambridge University Press.

Næss, A. (2008) *The Ecology of Wisdom: Writings by Arne Næss*, Berkeley, CA: Counterpoint.

Nash, R. (2001) *Wilderness and the American Mind* (fourth edition), New Haven and London: Yale Nota Bene.

Norton, B. (1995) 'Ecological integrity and social values: at what scale', *Ecosystem Health*, vol. 1, no. 4, pp. 228–241.

Noss, R. (1991) 'Sustainability and wilderness', *Conservation Biology*, vol. 5, pp. 120–123.

Oelschlaeger, M. (1991) *The Idea of Wilderness: from Prehistory to the Age of Ecology*, New Haven and London: Yale University Press.

Plumwood, V. (2001) 'Towards a progressive naturalism', *Capitalism, Nature, Socialism*, vol. 12, no. 4, pp. 3–32.

Plumwood, V. (2003) 'New nature or no nature?', unpublished paper provided by the author, 14 August 2003. This paper has since been incorporated in Plumwood, V. (2006) 'The concept of a cultural landscape: nature, culture and agency in the land', *Ethics and the Environment*, vol. 11, no. 2, pp. 115–150.

Reason, P. and Torbert, W. (2001) 'Toward a transformational social science: a further look at the scientific merits of action research', *Concepts and Transformations*, vol. 6, no. 1, pp. 1–37.

Rees, W. (2008) 'Toward sustainability with justice: are human nature and history on side?', in *Sustaining Life on Earth: Environmental and Human Health through Global Governance*, ed. C. Soskolne, New York: Lexington Books.

Rolston III, H. (1985) 'Valuing wildlands', *Environmental ethics*, vol. 7, pp. 23–48.

Rolston III, H. (1992) 'Disvalues in nature', *Monist*, vol. 75, no. 2, pp. 250–280.

Rolston III, H. (2001) 'Natural and unnatural: wild and cultural', *Western North American Naturalist*, vol. 61, no. 3, pp. 267–276.

Rolston III, H. (2012) *A New Environmental Ethics: the Next Millennium of Life on Earth*, London: Routledge.
Singer, P. (1981) *The Expanding Circle: Ethics and Sociobiology*, New York: Farrar, Straus & Giroux.
Smith, M. J. (1998) *Ecologism: Towards Ecological Citizenship*, Buckingham: Open University Press.
Soskolne C., Kotze, L., Mackey, B. and Rees, W. (2008) 'Conclusions: challenging our individual and collective thinking about sustainability', in *Sustaining Life on Earth: Environmental and Human Health through Global Governance*, ed. C. Soskolne, New York: Lexington Books.
Soulé, M. (1995) 'The social siege of nature', in *Reinventing Nature: Responses to Postmodern Deconstruction*, eds. M. Soulé and G. Lease, Washington, DC: Island Press, pp. 137–170.
Spring, J. (2004) *How Educational Ideologies are Shaping Global Society: Intergovernmental Organizations, NGO's, and the Decline of the State*, Mahwah, NJ: Lawrence Erlbaum.
Tacey, D. (2000) *Re-enchantment: the New Australian Spirituality*, Australia: Harper Collins.
Taylor, P. (1986) *Respect for Nature: a Theory of Environmental Ethics*, Princeton, NJ: Princeton University Press.
Tempest Williams, T. (1999) 'A place of humility', *Wild Earth*, vol. 9, no. 3, pp. 18–21.
'The intrinsic value of nature, special edition' (1992), *Monist*, vol. 75, no. 2.
Thomashow, M. (1996) 'Voices of ecological identity', in *Ecological Identity: Becoming a Reflective Environmentalist*, ed. M. Thomashow, Cambridge, MA: MIT Press.
Thoreau, H. D. (1995 [1854]) *Walden: or Life in the Woods*, New York: Dover Publications.
UNEP and UNESCO (1975) 'The Belgrade Charter', United Nations Environment Programme and United Nations Educational, Scientific and Cultural Organization, online, available at: http://unesdoc.unesco.org/images/0001/000177/017772eb.pdf (accessed 2 December 2014).
Washington, H. (2002) *A Sense of Wonder*, Sydney: Ecosolution Consulting.
Westra, R. (2004) 'The "impasse" debate and socialist development', in *New Socialisms: Futures beyond Globalisation*, eds. R. Albritton, S. Bell, J. Bell and R. Westra, London: Routledge.
Zack, N. (2002) 'Human values as a source for sustaining the environment', in *Just Ecological Integrity*, eds. P. Miller and L. Westra, Lantham, MD: Rowman and Littlefield, pp. 69–73.
Zerubavel, E. (2006) *The Elephant in the Room: Silence and Denial in Everyday Life*, London: Oxford University Press.
Zimmerman, M. (2001) *The Nature of Intrinsic Value*, Lanham, MD: Rowman and Littlefield.
Zohar, D. and Marshall, I. (2004) *Spiritual Capital: Wealth that we can Live By*, San Francisco: Berrett-Koehler.

9
AN UNSUSTAINABLE DENIAL

I seek to hide,
Insist that what 'is'
Will not be.
For I am afraid,
The change scares me,
So I defy it:
I proclaim it **must not be**.

(From 'Denial', Haydn Washington 2013a)

All of the environmental problems described in the preceding chapters are real. All of them have also been *denied*. Denied, not just by the odd crank, but by the majority of governments and 'we the people' over many decades. Not only do we deny the environmental crisis, we deny that we deny it. Laing (1971) notes that it is as if each of us were hypnotised twice: first into accepting pseudo-reality as reality, and second into believing we were not hypnotised. To demystify sustainability, we need to grasp how big a problem denial is. As Heidegger (1955) noted, society seems to be: 'in flight from thinking'. As a society, we continue to act as if there is no environmental crisis, no matter what the science says. As Rees (2010) concludes 'Modern society has been paralysed by deep-seated cognitive dissonance, collective denial, and political inertia in dealing with the unsustainability conundrum'.

More recently, we seem to speak of becoming 'sustainable', while at the same time denying the need to change. Hence it remains common for many to speak of a 'sustainable development' that is actually based on endless growth – the key driver of the environmental crisis. How is it possible for civilisations to be blind toward the grave approaching threats to their security, even when available evidence is accumulating about these threats? Existing laws have failed to deal with climate change, loss of biodiversity, air and water pollution, soil erosion, alien species, groundwater extraction, toxic substances and overexploitation of fisheries (Brown 2008). What on Earth is going on? Sadly, denial is as old as humanity, and possibly nobody is free from it (Zerubavel 2006). Poet T. S. Elliot in 'Burnt Norton' notes that 'Human kind ... cannot bear very much reality'. We proceed often in a cultural trance of denial, which Lifton and Falk (1982) refer to as 'psychic numbing', where people and societies block awareness of issues too painful to comprehend. This human incapacity to hear bad news makes it hard to solve the environmental crisis (Sukhdev 2010) or to reach sustainability.

Believing in stupid things

Rees (2010) notes that humanity is a 'deeply conflicted species' torn between what reason and moral judgment tell us on the one hand, and what baser emotions tell us to do on the other, particularly in stressful circumstances. We call ourselves *Homo sapiens*, but as we speak of denial, it will become apparent that many of us are really *Homo denialensis*. Yet society must deal with denial to have any chance of becoming 'sustainable' in any meaningful way. Humanity is very good at believing in *stupid things*, things that hold us back from reaching sustainability, such as:

1   the world and the Universe are all about *us*;
2   although we live on a finite planet, endless growth is possible (indeed apparently praiseworthy);
3   population growth is not a problem ('more is better');
4   endless growth in consumption and resource use is not a problem ('resource limits are in our mind');
5   the 'invisible hand' of the market is a god and must not be regulated;
6   technology can solve everything (techno-centrism);
7   greed is good.

The first is essentially anthropocentric and egotistical hubris; points two through four violate ecological reality and scientific laws (as well as common sense); the fifth is actually a religious myth (without scientific or ethical basis); the sixth is a modernist ideology that ignores ecological limits as well as ethics; while the seventh is unethical at so many levels, as noted by virtually every religious leader in history. Yet all of these are still espoused by various people in modern society. Indeed, many of them are still held by governments, the mass media, and by many of the public. Braungart and McDonough (2008, p. 117) note that insanity is defined as doing the same thing over and over and expecting a different outcome. The environmental crisis tells us that belief in the above points has failed, yet we keep doing them. The fact that many of us still believe these stupid things, and continue to assert that they are true (or even 'good') shows why we really must discuss *denial*. Denial is arguably the major barrier to us solving the environmental crisis and reaching a sustainable future. If we could abandon the belief in these delusional things, then the path to sustainability would become immensely easier.

Scepticism vs denial

But what is denial? Is it the same as scepticism, as some people seem to think? This is a crucial question for sustainability. The Oxford English Dictionary definition of a sceptic is: 'A seeker after truth; an inquirer who has not yet arrived at definite conclusions'. So we should all be sceptics in many ways, as we should all seek the truth. Scepticism is about seeking the truth and realising the world is a complex place. Genuine scepticism in science is one of the ways that science progresses, examining assumptions and conclusions (Pittock 2009). 'Denial' is something very different, a refusal to believe something no matter what the evidence. Those in denial demonstrate a 'wilful ignorance' and invoke logical fallacies to buttress their unshakeable beliefs (Specter 2009). In fact deniers commonly use a method of argument known as 'Bulverism' that assumes someone is wrong, then explains why they hold such a fallacious view (Hamilton 2010). One example would be: 'climate scientists are just wrong', and 'they are wrong because they are too liberal'. Denial isn't about searching for truth, it's about the denial of a truth one doesn't like. So scepticism and denial are in many ways opposites. An objective scientist should be sceptical, one should not jump to conclusions or believe something simply because it is fashionable. When sociologist Merton (1973) wrote about the structure of science, he stated that scientists operated by four principles: organised scepticism, universalism, communalism and disinterestedness. In the scientific method, true sceptics need to do three things: apply critical faculties to both sides of an argument, admit uncertainties on both sides of the argument, and accept that risk management may require appropriate policy responses despite the uncertainty (Pittock 2009). Deniers don't do these three things, while genuine 'sceptics' do. Scepticism is healthy in both science and society – denial is not.

It is thus important to understand the difference between scepticism and denial. Refusing to accept the overwhelming 'preponderance of evidence' is not scepticism, it is 'denial' and should be called by its true name. Accordingly, I speak here of *deniers* not sceptics. Others use the term 'denialists' but 'deniers' is more succinct and also acknowledges how common it is. In the United States the term 'contrarians' is also commonly used for climate change deniers (Hansen 2009).

Denial is common

How common is denial? We all deny, and the ability to deny is 'an amazing human phenomenon, largely unexplained and often inexplicable ... a product of the sheer complexity of our emotional, linguistic, moral and intellectual lives' (Cohen 2001). But what is it that we deny? Many things, the things we don't want to admit exist. In our daily life most of us tend to deny something, whether it's our looks, age, finances, or health. We deny some things as they force us to 'confront change', others because they are just too painful, or make us afraid. Sometimes we can't see a solution, so problems appear unsolvable. Thus many of us deny the root cause of the problem. After all, it would be much easier for us if the world would just go along in the same old comfortable way. Psycho-analysis sees denial as an 'unconscious defence mechanism for coping with guilt, anxiety or other disturbing emotions aroused by reality' (Cohen 2001). Sociologist Zerubavel (2006) in 'The Elephant in the Room' explains that the most public form of denial is 'silence', where some things are not spoken of. The silence about the environmental crisis, the silence about the fact that humans are dependent on Nature, the silence (until recently) about our dependence on ecosystem services, the deafening silence about the impossibility of endless growth. The apparent silence about how sustainability *requires major change*. All these silences are part of denial.

Zerubavel (2006) cites the fable of the 'Emperor's New Clothes' as a classic example of a conspiracy of silence, a situation where everyone refuses to acknowledge an obvious truth. He sheds light on the keeping of 'open secrets', and shows that conspiracies of silence exist at 'every level of society', ranging from small groups to large corporations, from personal friendships to politics. Such conspiracies evolve through the social pressures that cause people to deny what is right before their eyes. Each conspirator's denial is complemented by the others, and the silence is usually more intense when more people conspire, and especially when there are significant power differences amongst them (Zerubavel 2006). One key academic silence is that about the issue that on a finite planet, a meaningful 'sustainability' cannot be equated with 'sustainable development' based on endless growth (see Chapter 3).

Hence denial is common, and we need to understand this. As a form of denial, silence helps us avoid pain. When facing a frightening situation we often resort to denial. During the Second World War, early reports of Nazi massacres of Jews were actually dismissed by many Jews in Europe as sheer lies (Zerubavel 2006). The prospect of the 'final solution' was just too frightening to believe. Zerubavel notes that 'silence like a cancer grows over time', so that a society can collectively ignore 'its leader's incompetence, glaring atrocities and impending environmental disasters'. He concludes that denial is inherently *delusional* and inevitably distorts one's sense of reality.

People often get upset when confronted with information challenging their self-delusional view of the world. Many people prefer illusions to painful realities, and thus cherish their 'right to be an ostrich'. US Senator Moynihan has noted that 'everyone is entitled to his own opinion, but not to his own facts' (Gilding 2011). Tavris and Aronson (2007) point out the problem of 'cognitive dissonance', which is the rather sick feeling you get when you realise a cherished belief is not supported by the evidence. They explain that most people will go to great lengths to reduce cognitive dissonance. Quite commonly, that means denying the evidence that caused the dissonance. Hence people can deny scientific evidence that shows there is an environmental crisis. British politician Lord Molson famously stated 'I will look at any additional evidence to confirm the opinion to which I have already come' (Tavris and Aronson 2007). However, Zerubavel (2006) notes that the longer we ignore 'elephants' the larger they loom in our minds, as each avoidance triggers an even greater spiral of denial. The environmental crisis has now got to the point where the elephant is all but filling the room. We may now at times 'talk about' it, but we still deny it. Similarly, we talk about 'sustainability', but often deny the need for major changes.

It can be seen from the above that denial is everywhere. It is a fundamental part of the human psyche. Denial is however a delusion, one that can become a 'pathology' when it endangers the ecosystems that humans rely on. So denial is an understandable and very human trait, but a dangerous one. Just as the parable from historian Pliny the Elder of the ostrich sticking its head in the sand doesn't work for the ostrich, neither does denial of something that could be life-threatening. Ignoring a serious disease can lead to one's death. Similarly, denial of serious environmental problems may lead to the collapse of ecosystems upon which humanity depends (Washington 2013b). As Catton (1982) notes: 'But believing crash can't happen to us is one reason why it will'. Biologist Jared Diamond (2005) in 'Collapse' shows that societies that deny or ignore their problems do indeed collapse. He also notes that our modern society has more drivers pushing us toward collapse than ancient societies. Denial of some problems can thus be not only life-threatening, but even society-threatening. Probably the greatest denial today is about the problems of the growth economy (see Chapter 4). Next is climate change (or possibly population), a denial of something we find deeply worrying. The ramifications of this are huge, including a changed climate where many species are likely to go extinct, sea levels rise (flooding cities and populated coastal zones), worse droughts and desertification, and a world where agriculture will be hit hard, increasing famine (Pittock 2009; Washington and Cook 2011; Brown 2012; IPCC 2014).

Existing research in environmental sociology and social psychology has previously emphasised the notion that 'information' is the limiting factor in the public's lack of response on environmental issues, known as the 'information deficit model' (Norgaard 2011). There is the sense that 'if people only knew' they would act differently, such as drive less, or 'rise up' and put pressure on the government (Halford and Sheehan 1991). However, many scholars now believe this explanation is no longer tenable. There are other barriers than a lack of scientific knowledge to changing the status of climate change in the minds of the public – psychological, emotional and behavioural barriers. We need to understand the complex 'cultural circuits' of science communication in which framing, language, imagery, marketing devices, and media norms all play their part (Hulme 2009).

The history of denial

Denial is as old as humanity. Paul and Anne Ehrlich (1998) tabulated the common themes of denial anti-science:

- environmental scientists ignore the abundant good news about the environment;

- population growth does not cause environmental damage and may even be beneficial;
- humanity is on the verge of abolishing hunger, food scarcity is a local or regional problem and is not indicative of overpopulation;
- natural resources are superabundant, if not infinite;
- there is no extinction crisis, so action is both uneconomic and unnecessary;
- global warming and acid rain are not serious threats;
- stratospheric ozone depletion is a hoax;
- the risks posed by toxic substances are vastly exaggerated;
- regulation is wrecking the economy.

These arguments are still being made in virtually the same format (e.g. Plimer 2009). It is important to realise that today's environmental denial follows a long history of denial. The campaign to deny the need to protect natural areas and wilderness was one of the first great denials, and continues today (Oelschlaeger 1991; Washington 2006). The organisation 'Wise Use' denies the need for national parks or wilderness, opposes environmental regulation and sees no need for constraints on the exploitation of any resource for short-term gain (Helvang 1994; Ehrlich and Ehrlich 1998). Rachel Carson's (1962) 'Silent Spring' raised the problems of synthetic pesticides. This generated the next great environmental denial. DDT was praised by many as a 'saviour' of humanity by reducing malarial mosquito numbers (indeed it still is by some, such as Plimer 2009). Carson was vilified by the pesticide industry but biological research has shown that she was absolutely right (see Chapter 2).

The 'denial industry' for climate change (largely funded by fossil fuel companies) has been detailed in Monbiot's (2006) 'Heat', Hoggan's (2009) 'Climate Cover Up', Oreskes and Conway's (2010) 'The Merchants of Doubt', Washington and Cook's (2011) 'Climate Change Denial: Heads in the Sand' and Norgaard's (2011) 'Living in Denial'. What is apparent from the history of denial is that there is far more involved than merely 'confusion' about the science. There has been a deliberate attempt to muddy the waters and confuse the public, so that action is delayed or even halted. The denial industry deliberately seeks to sow doubt and confuse the public about environmental problems (Oreskes and Conway 2010). Al Gore (2006) quotes a tobacco company (Brown and Williamson) memo from the 1960s: 'Doubt is our product, since it is the best means of competing with the "body of fact" that exists in the minds of the general public. It is also the means of establishing a controversy'.

The Ehrlichs (1998) have detailed how the 'anti-science' denial movement promotes seemingly 'authoritative' opinions in books and the media that 'greatly distort what is or isn't known by environmental scientists'. The Ehrlichs noted this 'brownlash' has produced a body of 'anti-science' that is a 'twisting of the findings of empirical science – to bolster a predetermined worldview and to support a political agenda'.

Deniers argue that environmental regulation has gone too far, a common conservative theme (Oreskes and Conway 2010). Ehrlich and Ehrlich (1998) noted that denial anti-science interferes with and prolongs the 'already difficult search for realistic and equitable solutions to the human predicament'. There also exists 'greenscamming', where groups are formed that masquerade about being concerned over the environment, but actually work against the interests implied in their names. Enting (2007) describes these simply as 'front organisations', while Hoggan (2009) labels them 'astroturf' groups. Greenscamming is what biologists would call 'aggressive mimicry' (Ehrlich and Ehrlich 1998). Examples of such sham greenscam groups are the National Wetland Coalition, the Alliance for Environment and Resources, the National Wilderness Institute, the American Council on Science and Health, and the Global Climate Coalition. A detailed list for climate change denial and greenscam groups can be found at the end of 'Climate Change Denial' (Washington and Cook 2011). US congressman George Miller stated that these groups were seeking to disguise their actual motives, which were driven by profits and greed (Ehrlich and Ehrlich 1998).

Denial emerges when scientific method provides information that challenges the status quo and confronts people with uncomfortable facts about the state of the Earth's environment. Evidence of severe global environment degradation is then dismissed as 'inconclusive', and the messenger declared an 'emotive scaremonger' (Mackey 2008). Greenscam groups often use code phrases such as 'sound science' and 'balance', words that suggest objectivity, when in fact their 'sound science' means science or views that are interpreted to support denial anti-science (Ehrlich and Ehrlich 1998). Orthodox science that proceeds by peer-review, if it disagrees with the denial anti-science, is labelled as 'junk science'. This is thus a total inversion of reality. Denial anti-science (with few exceptions) does not proceed through the usual peer-review science process. In fact it is not science at all, merely unsupported statements without proven scientific evidence. Public relation companies have long been involved in 'spin', in seeking to modify the public's view of reality, and this is certainly the case with the denial industry, as noted by Hoggan (2009) who worked for many years in that industry.

Climate change denial is merely the last in a long line of denial anti-science campaigns. This started with DDT, then continued with nuclear winter, tobacco, acid rain and the ozone hole. Today almost everyone (but not some deniers) accepts that CFCs are responsible for the hole in the ozone layer, yet only ten years ago this was a hot topic of denial. The reason why denial of this is no longer strongly promoted is most likely that no powerful industry currently funds such denial. The refrigeration industry found economic alternatives to CFCs and the world has moved on, and the ozone hole is now starting to close (Hulme 2009). Denial books on the environmental crisis continue to emerge. Jacques et al. (2008) explain that between 1972 and 2005 there were some 141 denial books published, of which 130 came from conservative 'think tanks'. In the 1990s, 56 denial books came out, 92 per cent linked to right wing foundations or think tanks (Oreskes and Conway 2010). Probably the best known denial book is by Lomborg (2001) 'The Skeptical Environmentalist'. The Danish Committee for Scientific Dishonesty concluded his book had fabricated data, selectively discarded results and misused statistical methods.

It concluded that Lomborg was 'out of his depth' in the field of science (Hoggan 2009).

When considering the history of denial, we should remember that science (as in all fields) has a broad selection of people: radical, conservative, rational and irrational. Remember also how widespread denial is in the human psyche. There will always be 'someone' with a science degree somewhere who will champion almost any cause, and sometimes they will have PhDs and can even be professors (we have a couple in Australia). Just because there is a professor of something denying the environmental crisis does not mean it is not true, it just means that that particular professor is in denial. This is why one must make use of the 'preponderance of evidence' in science, the collective view. It should also be remembered that having a degree (or several degrees) in another field of science does not make one an expert in (for example) climate change. Despite this, it is common practice for denial organisations to quote one particular scientist who is in denial.

Monbiot (2006) concludes about the climate change denial industry that:

> By dominating the media debate on climate change during seven or eight critical years in which urgent international talks should have been taking place, by constantly seeding doubt about the science just as it should have been most persuasive, they have justified the money their sponsors spent on them many times over. I think it is fair to say that the professional denial industry has delayed effective global action on climate change by several years.

Hoggan (2009) describes the denial industry as:

> A story of betrayal, a story of selfishness, greed and irresponsibility on an epic scale. In its darkest chapters it's a story of deceit, of poisoning public judgement – of an anti-democratic attack on our political structures and a strategic undermining of the journalistic watchdogs who keep our social institutions honest.

Ideological basis for denial

The organised denial industry has played a major part in slowing or stopping action on environmental problems. It has held back moves towards sustainability. Oreskes and Conway (2010) detail the support that conservative think tanks have given to denial and ask 'what is going on?'. The link that united the tobacco industry, conservative think tanks and a group of denial scientists is that they were *implacably opposed to regulation*. They saw regulation as the slippery slope to socialism. They felt that concern about environmental problems was questioning the ideology of laissez-faire economics and free market fundamentalism. These conservative bodies equate the free market with liberty, so for them:

Accepting that by-products of industrial civilization were irreparably damaging the global environment was to accept the reality of market failure. It was to acknowledge the limits of free market capitalism ... science was starting to show that certain kinds of liberties are not sustainable – like the liberty to pollute.

(Oreskes and Conway 2010)

So if science impacts on their view of 'liberty', if it shows that regulation of pollution is needed, then science had to be opposed and denied. The basis for much denial is thus *ideological*, where science and the environmental crisis are denied due to a conservative ideological hatred of regulation affecting the free market. It is of interest that sociology until recently has overlooked the organised conservative opposition to climate change action (McCright and Dunlap 2000). Sociology and society as a whole can no longer afford to continue this.

Psychological types of denial

However, there is more involved than just the denial industry. Cohen (2001) describes three varieties of denial:

> **Literal denial** – the assertion that something did not happen or is not true. For global warming, this form of denial is seen in the generation of counter-claims by oil and coal companies that climate change is not happening (Gelbspan 1997; McCright and Dunlap 2000).
>
> **Interpretive denial** – in which the facts themselves are not denied, but they are given a different interpretation. Euphemisms and technical jargon are used to dispute the meanings of events. For example, military generals speak of 'collateral damage' rather than the killing of citizens. Interpretive denial is what we commonly call 'spin', and is commonly used by governments and business.
>
> **Implicatory denial** – where what is denied are 'the psychological, political or moral implications.... Unlike literal or interpretive denial, knowledge itself is not at issue, but doing the "right" thing with the knowledge' (Cohen, 2001). This is not negation of information per se, rather it is a failure to incorporate this knowledge into everyday life, or transform it into social action. People have access to information, accept this information as true, yet for a variety of reasons, *choose to ignore it* (Norgaard 2011).

Categorisation of the three types of denial is useful when considering the broad range of denial. The first, 'literal denial' is a key strategy of the denial industry. The second, 'interpretive denial' is generally what we see from government and business. This is what Fromm (1976) described as how governments 'give the impression that the problems are recognised and something is being done to resolve them. Yet nothing of real importance happens'. The third type, 'implicatory denial', is the type most common in the public. Implicatory denial is about 'bridging the moral and psychic gap between what you know and what you do' (Cohen 2001). Much of the knowledge about an environmental problem is accepted, but fails to be converted into action. We are vaguely aware of choosing not to look at the facts, but not quite conscious of just what it is we are evading (Cohen 2001). For example, the people in a study in Norway believed in climate change, expressed concern about it, yet lived their lives as though they 'did not know' (Norgaard 2011). Many of us could in fact be described as 'status apes', so if something is likely to threaten our status, we ignore or deny it.

'Distraction' is also an everyday form of denial. If we are worried about something, we tend to 'switch off' and shift our attention to something else. We can also 'de-problematise' it by rationalising that 'humanity has solved these sort of problems before' (Hamilton 2010). We can also 'distance ourselves' from the problem by rationalising 'it's a long way off'. There is also 'hairy-chested denial', where people deny climate change, as it will impact on pleasures such as big, fast cars. Blame-shifting is another part of implicatory denial, where we shift the blame onto others, such as the United States, China, industry, or the developing world (Hamilton 2010).

'Not knowing' certain things can be strategic (Norgaard 2011). To 'not know' too much about climate change maintains the notion of the 'innocence' of those involved. As Jensen (2000) observed 'for us to maintain our way of living we must … tell lies to each other, and especially to ourselves'. In 2004, a large survey of American citizens was carried out regarding climate change. They found that the 'more scientifically informed' respondents not only felt *less* personally responsible for global warming, but also showed less concern about it (Kellstedt *et al.* 2008). The 'information deficit model' is thus inadequate to understand how people respond to scientific controversies. Much more is involved in people's personal denial.

Zerubavel (2002) outlined the social dimension of 'ignoring' as the 'sociology of denial', where many people in society take part in co-denial. Norgaard (2003) noted that denial involves self-censoring or 'knowing what not to know'. By avoiding it we thus do nothing to solve the problems it represents. However, Zerubavel (2006) notes:

> 'Elephants' rarely go away just because we pretend not to notice them. Although 'everyone hopes that if we pretend not to acknowledge their existence, maybe … they will go away' even the proverbial ostrich that sticks its head in the sand does not really make problems disappear by simply wishing them away. Fundamentally delusional, denial may help keep us unaware of unpleasant things around us but it cannot ever actually make them go away.

The reason it is so difficult to talk about the elephant in the room is that 'not only does no one want to listen, but no one wants to talk about not listening' (Zerubavel 2006). We thus deny our denial. This is a particular form of self-deception made famous by Orwell (1949) in 'Nineteen Eighty-Four' as 'double-thinking'. Zerubavel (2006) notes that the longer we pretend not to notice the elephants, the larger they loom in our minds. However, one *can* break the silence of denial, and our strong need to deny things is counter-balanced by our equally strong need to expose denial. As soon as we acknowledge the elephant, it starts to shrink. Zerubavel (2006) explains that breaking conspiracies of silence implies 'foregrounding' the elephant in the room and calling a spade a spade. Denial can have its own tipping point, where 'increasing social pressure on the remaining conspirators to also acknowledge the elephant's presence eventually overrides the social pressure to keep denying it' (Zerubavel 2006). This is one tipping point we do need, to get society to acknowledge its denial of the environmental crisis.

Norgaard (2011) concludes that denial of climate change serves not only to manage negative emotions and preserve our sense of self-identity, but to maintain global economic interests and perpetuate global environmental inequalities. It is thus easier for many of us to slip into implicatory denial, rather than face the need for change. So why do we deny reality? Rees (2008) argues that in times of stress the brain's 'reptilian brain stem' (amygdala) will override the rational cortex, so we do stupid things. Thus our dedication to growth comes from these survival instincts. Rees (2008) concludes that if we are to survive, we must assert our capacity for 'consciousness, reasoned deliberation and willpower' to 're-write the "myths we live by" and articulate the necessary conditions for sustainability'. Rees (2010) expands on what he calls the 'triune' brain, where the reptilian brainstem and limbic system assume dominance over the rational neocortex. He raises the worrying issue that 'Intelligence and reason may not be the primary determinants of human behaviour at any social scale'.

Interestingly, a recent paper by Kanai *et al.* (2011) provides some evidence in support of Rees' conjecture about the amygdala. It found that greater political liberalism was associated with increased grey matter volume in the anterior cingulate cortex, whereas greater conservatism was associated with increased volume of the right amygdala. The amygdala has many functions (including fear processing), and individuals with a large amygdala are *more sensitive to fear*. They note also that one of the functions of the anterior cingulated cortex is to monitor uncertainty, and this may help people deal with an uncertain world. Sustainability means being able to deal with an uncertain world, not fear change to the extent we deny it is happening. Given that these differences in brain structure are most likely due to their early life in society, it shows the importance of this in terms of whether people are prone to denial or to finding solutions. The way people are taught, the society they grow up in, their connection to Nature, the worldview and ethics they are embedded in, these are all likely to shape how our brains develop. Hill (1992) notes: 'most of us have been significantly psychologically wounded as children and that the denial of this, and consequential lack of recovery, is the main barrier to our psychosocial development and to genuine progress towards sustainability'.

This brings us back to the importance of social sustainability (Chapter 6), and the fascinating question whether what has been called an 'enabling' society (Hill 2006), and how people live and are taught within this, may influence how their brains develop and whether they are more open to fear and denial or to solving our problems. It is thus possible that better social sustainability may decrease the key factors leading to denial.

Ways we let denial prosper

We should ask 'Do we let denial prosper?'. There are a number of aspects involved.

Fear of change: Of course many of us are not in denial, but it seems that many people and governments *are*. Why has there been a failure to feel an urgency about sustainability? One reason why many of us don't take action is the 'fear of change'. When struggling with the trauma of change, segments of society can turn away from reality 'in favour of a more comfortable lie' (Specter 2009). The fear of change is related to denial, and is a strong trait in humanity. Indeed, it is the hallmark of conservatism (though some prefer to call this neoliberalism or radical liberalism, Giddens 1994). Many conservatives just don't 'trust' scientists (and certainly not environmentalists) as they think they are too 'liberal' (and hence suspect). This was the conclusion of the Skeptical Science website (online, available at: www.skepticalscience.com), which examines climate change denial arguments. One climate change denier stated 'the cheerleaders for doing something about global warming seem to be largely the cheerleaders for many causes of which I disapprove' (Orthodoxnet 2007). Studies have consistently found conservatism to be negatively related to pro-environmental attitudes, especially among political elites (McCright and Dunlap 2000).

Conservatives see environmental regulation as threatening the core elements of conservatism, such as the primacy of individual freedom, private property rights, laissez-faire government, and promotion of free enterprise (McCright and Dunlap 2000). A binding treaty and carbon price are seen as a direct threat to sustained economic growth, the free market, national security and sovereignty, and the continued abolition of government regulations – all key conservative goals. Hoggan (2009) points out that in the United States, among highly educated people, 75 per cent of Democrats believe humans are causing climate change, but among Republicans it is only 19 per cent. McCright and Dunlap (2000) note that conservatives also tend to believe that 'radical environmentalists' are socialists 'who want to take over the world'.

Failure in worldview and ethics: Another reason why denial prospers is a failure in worldview, ethics and values. There is little discussion of worldview or ethics in modern society, as noted in Chapter 8. It is a truism that if you don't know where you are 'coming from' you don't know where you are going. If we don't actually strongly value the natural world that is in crisis and threatened, then we probably won't feel the urgency to save it.

Fixation on economics and society: As noted in Chapters 4, 5 and 6 our society has a fixation on economics and society. Our society looks first to economics and second to social issues. Hulme (2009) suggests that a confident belief in the human ability to control Nature is a dominant, if often subliminal, attribute of the international diplomacy that engages climate change. The intelligentsia in our society has mainly focused on social justice, and 'eco-justice' is rarely given equal time or prominence. Under these circumstances it is hard for many people to feel a sense of urgency, as ecological degradation is just not 'real' to them. This is partly due to the fact that our Western society is now more isolated from the natural world than ever before.

Ignorance of ecology and exponential growth: Another reason why we let denial prosper is our lack of an 'ecological grounding' (see Chapter 5). Most people don't understand how the world works in terms of ecosystems. The ecological grounding of our society has improved over the last 30 years, but many decision-makers are still woefully ignorant. Some of them still think environmental issues are just about 'tree hugging'. For this reason they feel no urgency to act on environmental issues. Another reason why people don't feel an urgency to act is our inability to really understand 'exponential growth' (Brown 1978). Many environmental problems are escalating at exponential rates, yet humans generally do not *think* exponentially. This is highlighted by the French riddle about a water lily in a pond, where the lilies double in size every day. On the thirtieth day the pond is full, so when is it half full? Many people reply 'the fifteenth day', when of course the answer is the twenty-ninth day (Brown 1978). Many environmental problems are worsening exponentially. Failure to understand exponential growth leads to a failure to act urgently on environmental problems, and aids denial.

Gambling on the future: Another aspect that delays us taking action is gambling on the future. People amazingly tend to discount the future (Giddens 2009). Many of us are gamblers at heart. If you are a gambler then you may not feel any sense of urgency, as you will always be gambling that things will turn out fine, or as we say in Australia: 'she'll be right mate!'.

The media and communication: Another aspect that assists denial to prosper is the media, which Ehrlich and Ornstein (2010) note has largely become a 'disinformation service'. The media poorly communicates environmental science. Hoggan (2009) explains that in Canada in 2006 the majority of people incorrectly blamed global warming on the ozone hole (due to poor media information). Hansen (2009) notes that 'scientific reticence' may hinder communication with the public. Scientists may be more worried about being accused of 'crying wolf' than they are about being accused of 'fiddling while Rome burns'. To put it another way, scientists are in a double-bind, for the demands of 'objectivity' suggest that they should keep aloof from contested issues, but if they do, then the public will not know what an objective view looks like (Oreskes and Conway 2010). Hansen (2009) asks whether future generations might not wonder how we today could have been 'so stupid as to do nothing'. He suggests that this would include scientists who did not adequately communicate the danger.

The media also loves *controversy*. This is often justified as being 'balanced' reporting, when in fact it shows major bias. A situation where the judgement of the vast majority of the scientific community is given equal space with denial advocates is anything but 'balanced'. In fact 'balance' has become a form of bias in favour of extreme minority views (Oreskes and Conway 2010). If one examines newspapers, one might think that science is split evenly about whether human-caused climate change is real, rather than more than 97 per cent of climate scientists saying this (Cook *et al.* 2013). Hoggan (2009) notes that often people from greenscam organisations (or think tanks) will be invited to write articles in newspapers, without acknowledging their organisation's links and funding. They are often presented as 'independent' scientists, when in fact often they are either not scientists, or come with a strong bias.

The media rarely checks factual accuracy. In 2007, the purported results of an 'important new scientific paper' were reported on the internet. It claimed that 'under-sea bacteria' were responsible for the build-up of CO_2 in the atmosphere. Within hours this supposed study was reported on denial blogs and circulating around the world through the media as 'evidence' to show that humanity did not cause climate change. Of course, it was a hoax created by journalist David Thorpe (see Hulme 2009). Thorpe did this to show that some people will use anything that supports their argument, whether it's true or not, and whether they checked its accuracy. These factors in the media make it easier for people to deny the environmental crisis, or at least ignore it. In fact, Oreskes and Conway (2010) argue it has made it easier for our governments to do almost nothing about climate change.

Conclusion

So clearly our society does let denial prosper. Rees (2008) concludes that on the dark side of myth, our shared illusions converge on deep denial. Our best science may tell us that the consumer society is on a self-destructive path, but we successfully deflect the evidence by repeating in unison the mantra of perpetual growth (Rees 2008). Westra (2008) believes the instrumental rationality of neoliberals (conservatives) is attuned to 'goals', not the concrete qualitative sustaining of life on Earth. What neoliberal policy has accomplished is a widening global inequality that exacerbates neglect of both population health and the global environment. Indeed, inequality fuels denial of environment concerns, as beneficiaries of neoliberal policy are protective and defensive of their gains (Westra 2008). 'Avoidance is sweet temptation' note Wackernagel and Rees (1996), but denial leads to 'greater pain tomorrow'. They conclude that the first step toward a more sustainable world is to 'accept ecological reality' and the socio-economic challenges it implies.

To conclude, denial is part of humanity, it has probably been with us since we first evolved. There are those who operate from greed and deliberately create a denial industry to confuse people. This is immoral and destructive. Similarly, there are the denial spin-doctors in government and business who seek to use 'weasel words' and fool the public that they are taking meaningful action, when often they just continue 'business-as-usual'. However, there is also denial in 'we the people'.

We let ourselves be duped and conned, we let our consciences be massaged, and we let our desire for the 'safe and easy life' blot out unpleasant realities. We delude ourselves. It is time to wake up, and break the denial dam. The key part of that dam regarding 'sustainability' is to understand that any meaningful 'sustainability' cannot be based on endless growth, even though the 'sustainable development' of 'Our Common Future' (WCED 1987) was based on this.

How big a problem is denial for sustainability? Like worldview, denial is central to sustainability. The environmental crisis is not unsolvable. However, you don't solve a problem you deny exists. The solutions exist and we discuss them in Chapter 11, but the solutions get harder the longer we deny the need for change. We could have acted three decades ago when Catton (1982) wrote 'Overshoot' and would now have been well on the path to a sustainable future. Instead, we denied we had a problem and deluded ourselves. Yet humanity *has* solved major problems, we acted on acid rain, we acted on CFCs, so we can act for sustainability. Humanity (when it chooses!) is intelligent and creative and can solve problems. The key issue now is to break the denial dam. Denial and sustainability are mutually exclusive, if you have one then you don't have the other.

Yet if a large part of the public abandons denial, they can fairly quickly turn around corporate denial, especially if it costs the corporations profits. If a large part of the community tells our politicians that they want real action (not weasel words), then politicians will actually *act*. We are not powerless drones who cannot change things. En masse, if we accept the 'Great Work' (Berry 1999) of repairing the Earth, we have the vision, the creativity and the power to solve the environmental crisis, and to move to a truly sustainable future. That nobody should deny.

References

Berry, T. (1999) *The Great Work*, New York: Belltower.
Braungart, M. and McDonough, W. (2008) *Cradle to Cradle: Remaking the Way we Make Things*, London: Vintage Books.
Brown, D. (2008) 'The ominous rise of ideological think tanks in environmental policy-making', in *Sustaining Life on Earth: Environmental and Human Health through Global Governance*, ed. C. Soskolne, New York: Rowman and Littlefield Publishers.
Brown, L. (1978) *The 29th Day*, Washington: Norton and Co.
Brown, L. (2012) *Full Planet, Empty Plates: the New Geopolitics of Food Scarcity*, New York: Norton.
Carson, R. (1962) *Silent Spring*, Boston: Houghton Mifflin.
Catton, W. (1982) *Overshoot: the Ecological Basis of Revolutionary Change*, Chicago: University of Illinois Press.
Cohen, S. (2001) *States of Denial: Knowing About Atrocities and Suffering*, Cambridge: Polity Press.
Cook, J., Nuccitelli, D., Green, S., Richardson, M., Winkler, B., Painting R., Way, R., Jacobs, P. and Skuce, A. (2013) 'Quantifying the consensus on anthropogenic global warming in the scientific literature', *Environmental Research Letters* vol. 8, no. 2, pp. 1–7, online, available at: http://iopscience.iop.org/1748-9326/8/2/024024/pdf/1748-9326_8_2_024024.pdf (accessed 1 September 2013).

Diamond, J. (2005) *Collapse: Why Societies Choose to Fail or Succeed*, New York: Viking Press.
Ehrlich, P. and Ehrlich, A. (1998) *Betrayal of Science and Reason: How Anti-environmental Rhetoric Threatens Our Future*, New York: Island Press.
Ehrlich, P. and Ornstein, R. (2010) *Humanity on a Tightrope: Thoughts on Empathy, Family and Big Changes for a Viable Future*, New York: Rowman and Littlefield.
Enting, I. (2007) *Twisted: the Distorted Mathematics of Greenhouse Denial*, Melbourne: Australasian Mathematical Sciences Institute.
Fromm, E. (1976) *To Have or to Be*, New York: Abacus Books.
Gelbspan, R. (1997) *The Heat is On: the High Stakes Battle Over Earth's Threatened Climate*, Reading, MA: Addison-Wesley Publishing Company.
Giddens, A. (1994) *Beyond Left and Right: the Future of Radical Politics*, Palo Alto, CA: Stanford University Press.
Gilding, P. (2011) *The Great Disruption: How the Climate Crisis will Transform the Global Economy*, London: Bloomsbury.
Gore, A. (2006) *An Inconvenient Truth*, New York: Bloomsbury Publishing.
Halford, G. and Sheehan, P. (1991) 'Human responses to environmental changes', *International Journal of Psychology*, vol. 269, no. 5, pp. 599–611.
Hamilton, C. (2010) *Requiem for a Species: Why We Resist the Truth about Climate Change*, Australia: Allen and Unwin.
Hansen, J. (2009) *Storms of my Grandchildren: the Truth about the Coming Climate Catastrophe and our Last Chance to Save Humanity*, London: Bloomsbury.
Heidegger, M. (1955) *Martin Heidegger: Philosophical and Political Writings*, New York: Continuum International.
Helvang, D. (1994) *The War Against the Greens: the 'Wise Use' Movement, the New Right, and Anti-environmental Violence*, San Francisco: Sierra Club Books.
Hill, S. B. (1992) 'Ethics, sustainability and healing', Talk to Alberta Round Table on the Environment and Economy and the Alberta Environmental Network, 11 June, online, available at: www.beliefinstitute.com/article/ethics-sustainability-and-healing (accessed 23 June 2014).
Hill, S. B. (2006), 'Enabling redesign for deep industrial ecology and personal values transformation: a social ecology perspective', in *Industrial Ecology and Spaces of Innovation*, eds. K. Green and S. Randles, Cheltenham, UK: Edward Elgar Publishing.
Hoggan, J. (2009) *Climate Cover Up: the Crusade to Deny Global Warming*, Vancouver: Greystone Books.
Hulme, M. (2009) *Why We Disagree About Climate Change: Understanding Controversy, Inaction and Opportunity*, Cambridge: Cambridge University Press.
IPCC (2014) Summary for Policymakers, Fifth Assessment Report, International Panel on Climate Change, online, available at: http://ipcc-wg2.gov/AR5/images/uploads/WG2AR5_SPM_FINAL.pdf (accessed 16 June 2014).
Jacques, P., Dunlap, R. and Freeman, M. (2008) 'The organisation of denial: conservative think tanks and environmental scepticism', *Environmental Politics*, vol. 17, no. 3, pp. 349–385.
Jensen, D. (2000) *A Language Older than Words*, New York: Context Books.
Kanai, R., Feilden, T., Firth, C. and Rees, G. (2011) 'Political orientations are correlated with brain structure in young adults', *Current Biology*, vol. 21, pp. 677–680.
Kellstedt, P., Zahran, S. and Vedlitz, A. (2008) 'Personal efficacy, the information environment, and attitudes towards global warming and climate change in the United States', *Risk Analysis*, vol. 28, no. 1, pp. 113–126.
Laing, R. D. (1971) *The Politics of the Family*, New York: Penguin.
Lifton, R. and Falk, R. (1982) *Indefensible Weapons*, New York: Basic Books.

Lomborg, B. (2001) *The Skeptical Environmentalist: Measuring the Real State of the World*, Cambridge: Cambridge University Press.
Mackey, B. (2008) 'The Earth Charter, ethics and global governance', in *Sustaining Life on Earth: Environmental and Human Health through Global Governance*, ed. C. Soskolne, New York: Rowman and Littlefield Publishers.
McCright, A. and Dunlap, R. (2000) 'Challenging global warming as a social problem: an analysis of the conservative movement's counter-claims', *Social Problems*, vol. 47, no. 4, pp. 499–522.
Merton, R. (1973) *The Sociology of Science: Theoretical and Empirical Investigations*, New York: University of Chicago Press.
Monbiot, G. (2006) *Heat: How to Stop the Planet Burning*, London: Penguin Books.
Norgaard, K. (2003) 'Denial, privilege and global environmental justice: the case of global climate change', paper presented at the annual meeting of the American Sociological Association, Atlanta Hilton Hotel, Atlanta, online, available at: www.allacademic.com.
Norgaard, K. (2011) *Living in Denial: Climate Change, Emotions, and Everyday Life*, Cambridge, MA: MIT Press.
Oelschlaeger, M. (1991) *The Idea of Wilderness: from Prehistory to the Age of Ecology*, New Haven and London: Yale University Press.
Oreskes, N. and Conway, M. (2010) *Merchants of Doubt: How a Handful of Scientists Obscured the Truth on Issues from Tobacco Smoke to Global Warming*, New York: Bloomsbury Press.
Orthodoxnet (2007) Post 14 by 'Augie' on 10 June 2007 at Orthodoxnet website, online, available at: www.orthodoxytoday.org/blog/2007/06/03/they-call-this-a-consensus/#comment-95231 (accessed 2 December 2014).
Orwell, G. (1949) *Nineteen Eighty-Four*, London: Secker and Warburg.
Pittock, A. B. (2009) *Climate Change: the Science, Impacts and Solutions*, Australia: CSIRO Publishing/Earthscan.
Plimer, I. (2009) *Heaven + Earth: Global Warming: The Missing Science*, Ballan, Vic: Connorcourt Publishing.
Rees, W. (2008) 'Toward sustainability with justice: are human nature and history on side?', in *Sustaining Life on Earth: Environmental and Human Health through Global Governance*, ed. C. Soskolne, New York: Lexington Books.
Rees, W. (2010) 'What's blocking sustainability? Human nature, cognition and denial, *Sustainability: Science, Practice and Policy*, vol. 6, no. 2 (ejournal), online, available at: http://sspp.proquest.com/archives/vol. 6iss2/1001–012.rees.html (accessed 21 June 2014).
Specter, M. (2009) *Denialism: How Irrational Thinking Hinders Scientific Progress, Harms the Planet and Threatens our Lives*, New York: The Penguin Press.
Sukhdev, P. (2010) 'Preface' to *The Economics of Ecosystems and Biodiversity: Ecological and Economic Foundations*, ed. P. Kumar, London: Earthscan.
Tavris, C. and Aronson, E. (2007) *Mistakes Were Made (But Not By Me): Why We Justify Foolish Beliefs, Bad Decisions, and Hurtful Acts*, Orlando, FL: Harcourt Books.
Wackernagel, M. and Rees, W. (1996) *Our Ecological Footprint: Reducing Human Impact on the Earth*, Gabriola Island, BC, Canada: New Society Publishers.
Washington, H. (2006) 'The wilderness knot', PhD Thesis, Sydney: University of Western Sydney, online, available at: http://arrow.uws.edu.au:8080/vital/access/manager/Repository/uws:44 (accessed 2 December 2014).
Washington, H. (2013a) *Poems from the Centre of the World*, online, available at: www.lulu.com/au/en/shop/haydn-washington/poems-from-the-centre-of-the-world/paperback/product-21255751.html (accessed 2 December 2014).
Washington, H. (2013b) *Human Dependence on Nature: How to Help Solve the Environmental Crisis*, London: Earthscan.

Washington, H. and Cook, J. (2011) *Climate Change Denial: Heads in the Sand*, London: Earthscan.

WCED (1987) *Our Common Future*, World Commission on Environment and Development, London: Oxford University Press.

Westra, R. (2008) 'Market society and ecological integrity: theory and practice', in *Sustaining Life on Earth: Environmental and Human Health through Global Governance*, ed. C. Soskolne, New York: Lexington Books.

Zerubavel, E. (2002) 'The elephant in the room: notes on the social organization of denial', in *Culture in Mind: Toward a Sociology of Culture and Cognition*, ed. K. Cerulo, New York: Routledge.

Zerubavel, E. (2006) *The Elephant in the Room: Silence and Denial in Everyday Life*, London: Oxford University Press.

10
APPROPRIATE TECHNOLOGY FOR SUSTAINABILITY

> ...the "fossil-nuclear metabolism" of industrial society has no future. The longer we cling to it the higher the prices will be for future generations.
> (German Advisory Council on Global Change, WBGU 2011, p. 25)

Appropriate technology

We cannot leave sustainability without discussing the part technology should play in a truly sustainable future. The techno-centrists and Cornucopians see technology as a god that will solve all problems. However, President Obama's chief science advisor, John Holdren, noted in 2007 that 'belief in technological miracles is generally a mistake' (Nicholson 2013, p. 317). Meadows et al. (2004, p. 206) note that technological advance, or markets by themselves, unchanged, unguided by understanding, respect, or commitment to sustainability, cannot create a sustainable society. Others again see technology as the *cause* of the environmental crisis, that it is inherently flawed, as being part of a 'scientism' (Daly 1991) that is destroying the Earth. Who is right? 'Technology' is generally lumped together as just one thing, when in fact it has many parts and strands. It is not uniform and consistent. Some parts can be key solutions to the environmental crisis and reaching sustainability. Some are not. It has promising aspects and very worrying ones. Chapter 8 has shown that the 'mastery of Nature' aspect of some science and technology has led us to disaster. But this is an 'ideology' pasted onto science, not something inherent in either science or technology. Science and technology do not have to be this way. Part of the problem has not been science or even technology per se, but that they have not been *appropriate*. Appropriate to what? Appropriate to the reality we face: an environmental crisis. Appropriate to the rapidly reducing time-frame over which we need solutions. Appropriate for a sustainability worldview and an 'Earth ethics' (Rolston 2012). Schumacher (1973) called for 'appropriate' technology to lead us to sustainability. We need an environmental revolution, but unlike the Industrial Revolution, this one must use appropriate technology, but will also be driven by our need to 'make peace with Nature' (Brown 2006).

So in terms of appropriate technology – are there solutions? Yes! This is one of the most frustrating aspects of the whole debate, the denial of real and workable solutions. The environmental crisis is not an irrevocable 'Decree of Fate' to which we must submit. It is a human-caused problem and has human-invented solutions, *provided we act*. One of the most frustrating aspects is that appropriate technology (such as renewable energy) will create a large number of green jobs, more jobs than the fossil fuel and mining industries combined. Yet many who supposedly support 'job creation' (e.g. conservatives) will (in the same breath) oppose renewable energy.

Key amongst these appropriate technologies is *renewable energy*. The Sun radiates to Earth each year around 6,600 times (Diesendorf 2014) as much energy as humans currently consume. There is thus no problem with there being enough renewable energy 'available'. The issue has always been harnessing this effectively and economically. The renewable energy sources available are all available essentially forever (in terms of human history). Why then has there been a history of ignoring, denying or even repressing renewable energy? The answer is simple: oil and coal. Society went down the coal path in the industrial revolution, and became addicted to coal for stationary energy use and steam trains. Then oil was discovered, and provided an easy liquid fuel for transport. We became addicted to both. They make great wealth for corporations, who will fight to continue this until they are stopped in the public interest. We need alternatives to both. Chapter 9 explains the intense denial of climate change, fostered by the fossil fuel denial industry. Renewable energy however suffers from its own negative myths and denial, as we shall see.

The alternatives exist to a fossil fuel based system, both in terms of technologies and the policies for implementing them. I will not list them in great detail, as others have already done this for technologies (e.g. Jacobson and Delucchi 2011; Boyle 2012; Diesendorf 2014) and policies (e.g. WBGU 2011; GEA 2012). Before discussing particular renewable energy technologies, we should discuss a common dismissal of renewable energy, which is the statement: 'The Sun doesn't shine at night!'. In fact, concentrated solar thermal (CST) with a heat sink does run at night, as do wind and geothermal. These all have high reliability, and wind, solar and some geothermal are proven technologies (Diesendorf 2013a). In fact coal power stations are quite inflexible, they take time to start and stop, can break down unexpectedly and their output cannot be varied rapidly enough to meet the peaks in demand. It is this very inflexibility that the coal industry has sought to turn into a virtue under the title 'baseload' power. What we should consider is the actual 'reliability' of the system overall, rather than whether each renewable technology can provide 'baseload' (AWEA 2012). The mix of renewables discussed below can be just as reliable as current coal-fired generation (Elliston *et al.* 2012, 2014), in fact its flexibility means it may be more reliable. The concept of 'baseload' power thus becomes redundant.

Renewable energy supplied an estimated 19 per cent of global final energy consumption at the end of 2012 (REN21 2013). In regard to electricity, renewable energy generating capacity reached 1,470 GW (440 GW excluding hydroelectricity) of electricity in 2012. At the end of 2012, renewables provided 26 per cent of electricity generation capacity, and an estimated 21.7 per cent of annual global electricity, with 16.5 per cent of electricity provided by hydropower (REN21 2013). By 2012, 138 countries had renewable energy targets (REN21 2013). Total investment in renewable energy reached $244 billion in 2012 (REN21 2013). Denmark plans to reach 100 per cent renewable electricity by 2035, Scotland by 2020 and New Zealand plans to reach 90 per cent by 2025 (Diesendorf 2014). Solar photovoltaic (PV) costs may fall to 6–10 c/kWh by 2020 (IRENA 2012, p. 38). However the more accepted figure for PV and CST is that they will fall to 10–12 c/kWh in the medium-term (Diesendorf 2014). Such figures are much cheaper than nuclear power, and also cheaper than coal-fired electricity in many areas (Klein 2010).

Some studies on renewable energy are worth special mention, such as the 2010 study 'Zero Carbon Britain 2030' (Kemp and Wexler 2010) from the Centre for Alternative Technology in Wales. Based on a mix of technologies that are either commercially available or at the demonstration stage, it describes in detail a feasible renewable energy vision for the United Kingdom in 2030 with 'no net GHG emissions'. The detailed Global Energy Assessment (GEA 2012) studied 60 future scenarios for renewable energy, and found that 41 fulfilled all the goals set for the study, showing there are many ways of providing a technically feasible, affordable, environmentally sustainable and socially equitable energy future, compatible with available resources.

There are many 'myths' about renewable energy, and these are analysed and answered in the illuminating book by Mark Diesendorf (2014) 'Sustainable Energy Solutions for Climate Change'. The three main myths are: (1) It is too diffuse for an industrial civilisation; (2) It is too intermittent and unreliable; and (3) It is too expensive to be used on a large scale. Diesendorf (2014) thoroughly debunks each of these myths. Most studies and scenarios suggest we already have most of the technologies needed to dramatically reduce greenhouse gas emissions through energy efficiency, demand reduction and renewable energy. There are also studies showing that the transition, although initially expensive in terms of investment, is affordable. The principal barriers are the absence of effective policies to rapidly drive the transition and this situation results from the political power of the fossil fuel industry (Diesendorf 2014).

Civilisation *can* reach a 95 per cent sustainably-sourced energy supply by 2050. There are however up front investments required to make this transition in the coming decades (1–2 per cent of global GDP), but they will turn into a positive cash flow after 2035, leading to a positive annual result of 2 per cent of GDP in 2050 (WWF 2011). A large-scale wind, water and solar energy system can reliably supply all of the world's energy needs, with significant benefit to climate, the environment, and energy security, at reasonable cost (Delucchi and Jacobson 2011).

However, energy efficiency and conservation must also play a key role in any sustainable energy future (WWF 2011; IEA 2012, p. 297). In reality, world energy demand *cannot keep on growing*. World energy use has grown exponentially by around 3 per cent a year for the past two centuries (USEIA 2011). If it was to continue at this rate, our current rate of energy use of 16 TW would balloon out to equal the 'entire solar output' in 1,000 years, and match all the energy output of the 100 billion stars in our galaxy inside of 2,000 years. Well before this, in just 400 years, enough heat would be generated to bring the Earth's surface temperature to that of boiling water. Obviously the so-called 'normal' world of energy growth is a temporary anomaly, destined to self-terminate by natural means (Murphy 2013, p. 173). One key task of 'sustainability' clearly is to halt exponential energy growth and reduce total energy use, and then supply this through appropriate technology.

Our current profligate energy use has already caused the environmental crisis. Overall we don't need 'more' energy (as claimed by technocentrists and nuclear advocates), we need *less*. Within this reality, the developed world in particular must reduce energy use, while the developing world increases somewhat for equity reasons. If we halve our energy use through energy conservation and energy efficiency (and strategies such as 'Factor Five', von Weizsäcker *et al.* 2009), then it is perfectly feasible to supply this through renewable energy. Rather than continuing to grow each year, energy use then needs to decline in line with the reduction of our unsustainable human population, and unsustainable consumption (see Chapter 7).

Renewable energy technologies

Renewable energy technologies are covered in detail in Boyle (2012) and Diesendorf (2014), but the main ones are summarised below:

- *Windpower*. The cost of wind energy has come down hugely over the last 30 years. It is already cheaper than nuclear electricity and, in some locations, around par with coal-fired and gas-fired electricity in cost (BNEF 2013; Diesendorf 2014). Over the past ten years, global wind power capacity has continued to grow at an average cumulative rate of over 30 per cent, with a total installed capacity of 282.4 GW at the end of 2012 (Diesendorf 2014). It is forecast to reach 450 GW by 2015 (GWEC 2010). The Global Wind Energy Council estimates that wind energy could supply up to 12 per cent of global electricity by 2020, and more than 20 per cent by 2030, provided it is backed by effective policies in governments (GWEC 2010). The wind doesn't always blow in any one spot, but it does blow 'somewhere' most of the time. A widely distributed wind power network will thus form part of a reliable renewable energy system (Diesendorf 2007). Several countries met high shares of their electricity demand with wind power in 2012, including Denmark (30 per cent), Portugal (20 per cent), Spain (16.3 per cent), and Ireland (12.7 per cent) (REN21 2013). Noise and bird death are over-rated problems with wind power. Bird deaths caused by wind turbines are a minute fraction of the total human-caused bird deaths: less than 0.003 per cent in 2003 (NRC 2007).

- *Solar.* Solar thermal is ideal for water heating and can make a significant contribution to space heating and cooling. Electricity from the Sun can be produced either by concentrated solar thermal (also called concentrated solar thermal power, CSP) or by solar photovoltaic. At the end of 2012 there were 2.5 GW of CST installed worldwide (REN21 2013), and 15 GW are under development or construction (IEA 2010). Recently, growth in CST has slowed as the cost of PV decreases (REN21 2013). A proposal is under consideration for a vast solar thermal installation in the Sahara that could provide a sixth of Europe's electricity needs (Kanter 2009; see Desertec Foundation's website, online, available at: www.desertec.org/en/concept/). Solar thermal storage systems (such as molten salts of sodium and potassium nitrate) can store solar energy for 15 hours or more for later use at night. In this way solar thermal can add to the reliability of renewable systems. Estimates indicate CST could reach 6,000 GW by 2050 (WWF 2011). By 2050, with appropriate support, CST could provide 11.3 per cent of global electricity, with 9.6 per cent from solar power and 1.7 per cent from backup fuels (fossil fuels or biomass) (IEA 2010).

  At the end of 2012 there was 102 GW of PV installed globally, with 31.1 GW being installed in just 2012 (EPIA 2013). Annual growth rates are in the order of 25–30 per cent (WWF 2011). In 2012 in the European Union, almost 70 per cent of electricity generation installed that year came from renewables (REN21 2013). The prospects for rapid growth of solar (thermal and PV) are excellent. This is contrary to the views expressed by nuclear and denial advocates. The two main central problems of solar – remoteness from markets and the need for flexible back-up or storage to counter its intermittent nature – have now been largely solved (Diesendorf 2014).

- *Geothermal and hot rocks.* Geothermal traditionally harnesses the steam produced naturally by the Earth's heat. Global generating capacity in 2011 was 11.7 GW (REN21 2013), almost all being conventional geothermal, where hot water reaches the Earth's surface in volcanic regions. However, conventional geothermal is limited to just a few regions of the world. There is far more potential energy in non-volcanic regions stored in hot rocks 3–5 km below the surface. Such 'hot rocks' can be used to produce steam if water is injected, which is then run through a steam turbine. Hot rock geothermal technology has a few small-scale pilot plants scattered around the world and is claimed to be on the brink of medium-scale demonstration in Australia (Geodynamics 2011).

- *Hydroelectricity.* Hydroelectricity is currently the most developed form of renewable electricity, producing 990 GW at the end of 2012 (REN21 2013). This still has large potential in Africa and parts of Asia. However, large hydroelectricity dams (such as the Three Gorges in China and Sardar Sarova in India) generally have major adverse environmental and social impacts (IR 1996). Small and medium scale hydro projects in widely distributed areas are likely to

be better options, though these too can have issues (e.g. siltation and eutrophication) that need to be considered (Makhijani and Ochs 2013).
- *Wave and tidal power (and ocean currents)*. The power of the tides has been harnessed in a few places in the world (e.g. the Rance Estuary in France) with around 527 MW in operation (REN21 2013). However, conventional tidal power is limited to a few regions of the world where very high tides prevail. Wave power has been harder to harness, but there are demonstration plants in Portugal, the United States, Germany, Scotland and Australia (REN21 2013).
- *Bioenergy*. Globally, photosynthesis stores solar energy in biomass (organic material) at a rate of about seven times the 2010 rate of global energy use (GEA 2012). So there is a substantial biomass resource world-wide. Plants store solar energy in wood, fibre and oils. These can be burned to produce energy for both stationary sources and transport, or converted to liquid and gaseous fuels by a wide range of processes. Bioenergy is the oldest form of renewable energy. Unless there is major transport involved (using fossil fuels), or inputs of fertilizers produced with fossil fuels, then such bioenergy comes close to being 'carbon neutral'. Plants take up CO_2, which is then released when they are burnt. Diesendorf (2007) notes that bioenergy offers both threats and opportunities. If poorly implemented it can result in substantial greenhouse gas emissions, in soil depletion, consumption of scarce water resources, and loss of biodiversity and old growth forests. If implemented appropriately, it could have very low environmental impacts, could help restore degraded land, and could play an important part in a sustainable renewable future (WWF 2011).

However, confusion over biofuels is rife. They can be useful in reducing carbon emissions (especially regionally), if sourced from plantations or agricultural residues. Yet they can be harmful if you clear existing rainforest to grow oil palms (as in Indonesia), or clear old growth forest to burn in power stations. However, environmentally unsound practices, such as the production of ethanol from corn in the United States, and the clearing of tropical forests to grow palm oil trees in southeast Asia, need not blind us to the possibilities of using biomass on a limited scale and in environmentally sustainable ways (Diesendorf 2014). Bioenergy must demonstrate a significant *net* carbon reduction over its whole life cycle. It should never come from native forests, but from mixed-species plantations or agricultural and sawmill wastes. Bioenergy electricity plants (biogas turbines and biodiesel generators) could be used to provide peak load demand in an integrated renewable energy system (Elliston *et al.* 2014).

Bioenergy is a significant but over-looked source of renewable energy. Biomass is also important in terms of replacing 'energy-dense' construction materials. Wood has an energy density of 1–3 MJ/kg whereas steel is 34 and aluminium is 170 (Pittock 2009). Bioenergy can also be obtained from the production of 'biochar' (charcoal), where the biochar is then used as an agricultural product that sequesters carbon into soil, and also improves water and nutrient retention. Biochar is not just carbon neutral, but *carbon-negative*, removing carbon from the atmosphere and putting it in soils (Lehmann *et al.* 2006; Taylor 2010). This is one of the most important carbon-negative technologies available, given that climatically we have already overshot the 'safe' level of CO_2 in the atmosphere (Washington and Cook 2011). Estimates vary of how much carbon biochar can sequester in soils. The International Biochar Initiative (IBI 2012) estimates that we could achieve 1 Gt of carbon sequestered annually before 2050. Lehmann *et al.* (2006) argue that by 2100 we could achieve 5.5–9.5 Gt/yr carbon sequestered. The most rigorous estimate is that biochar could sequester up to 1.8 Gt/yr, or 12 per cent of human CO_2 emissions (Woolf *et al.* 2010). The potential of biochar to lower atmospheric CO_2 is thus significant. It is a pity that it has not received the attention from governments it deserves.

The use of all these technologies (often suited to remote locations) has been improved by the ability to use high-voltage DC transmission cables with emission losses of less than 3 per cent over 1,000 km (which is 30–40 per cent less than for normal AC transmission lines, Siemens 2010).

To solve climate change will require a rapid and major conversion to energy efficiency and renewable energy, as we have delayed for so long. We need an urgent large-scale programme. This major effort in renewable energy is a key part of the 'Great Work' of sustainability. The money needed would be much less than what we already spend on the military (Pittock 2009). In Australia we currently subsidise fossil fuels to the tune of $10 billion a year (Riedy and Diesendorf 2003). If this was directed to renewable energy, it would likely be enough to rapidly transition Australia to renewable energy within two to three decades (Diesendorf 2013a). A recent study in Australia has claimed that Australia could become renewable-powered in ten years at a cost of 3 per cent of GDP a year (Wright and Hearps 2010). The assumptions used, the energy source split, the modelling and the feasibility of the time-span have been questioned (Diesendorf 2011), though not the fact that a shift to 100 per cent renewables *can* be done (Diesendorf 2014). No energy technology is entirely benign. Nevertheless, energy efficiency and most renewable energy sources (that are sited sensibly) are essentially everlasting and safe, have very low environmental and health impacts, can create many local jobs and have rapid energy payback periods. In terms of these criteria they are far superior to fossil and nuclear energy (Diesendorf 2014). Renewable energy is thus entirely feasible and sustainable.

However, unless we reduce rampant consumerism, the sheer scale of renewables required to supply an increasingly affluent world in the time required would be mind-boggling (Assadourian 2010). We thus need a renewable change-over that goes hand in hand with a reduction in total energy demand, at least in developed countries, and a reduction in consumerism everywhere. For years we buried our heads in the sand and denied the potential of renewable energy. However, the technologies exist, are feasible and affordable – all we need is the *political will*.

Inappropriate technologies

Some *inappropriate* technologies need brief discussion. The first is *nuclear power*. It is not the answer because you don't solve one serious problem with another. Energy expert Amory Lovins (2006) asks, 'Why keep on distorting markets and biasing choices to divert scarce resources from the winners to the loser – a far slower, costlier, harder, and riskier niche product – and paying a premium to incur its many problems?'.

So why is society debating its merits? The reason is the powerful nuclear and uranium mining lobbies, which (due to concern over climate change) see a chance to resurrect their nuclear fantasy. Nuclear power is *not* a viable solution because:

- It's *not carbon neutral anyway*, and can produce 10–40 per cent as much CO_2 as a coal-fired power station, depending on the uranium ore grade (Mudd and Diesendorf 2008). Nuclear energy results in 9–25 times (Jacobson and Delucchi 2011) or 6–22 times (Diesendorf 2014) more carbon emissions than wind energy, depending on the grade of the uranium ore. With low grade ore, the CO_2 emissions are comparable with those from natural gas. Hence nuclear energy, based on existing commercial technology, cannot be a long-term energy/electricity solution to global climate change (Diesendorf 2014).
- Nuclear energy is *too slow to deploy*. It takes many years to commission a nuclear power station. The historic 'planning-to-operation' times for new nuclear plants has been 11–19 years, compared with an average of 2–5 years for wind and solar installations (Jacobson and Delucchi 2011). Wind farms could generate the same power for 60 per cent of the construction costs, and are much cheaper to run (Lovins and Sheikh 2008).
- Nuclear power is *not cheap*. The current cost of nuclear is estimated to be at least 8–12 c/kWh (Jacobson and Delucchi 2011) or more likely 17–34 c/kWh (Diesendorf 2014). Delucchi and Jacobson (2011) state the cost of wind energy is around 4–7 c/kWh and solar CST is 11–15 c/kWhr, which should drop to 8 cents by 2020. Diesendorf (2014) lists the cost of wind energy as 8–15 c/kWh, and solar PV and CST (parabolic trough) as 26 c/kWh. Nuclear is thus much more expensive than wind power, and about par with large-scale solar, though the cost of solar is dropping, while nuclear costs are likely to keep rising. If the huge subsidies to nuclear power are taken into account (Diesendorf 2013b) and the hidden external costs (Sovacool 2011), nuclear is even more expensive than solar electricity (Diesendorf 2014).
- Conventional nuclear fission relies on *finite reserves of high-grade uranium*. A large-scale conventional nuclear programme would exhaust these in less than a century (Jacobson and Delucchi 2011). There are not enough total uranium reserves to provide a low-carbon future anyway, *unless* we move to 'fast breeder' reactors. These have major safety problems with potential for a large (non-nuclear) explosion that could disperse radioactive material into the environment (Kumar and Ramana 2009). By creating much larger quantities

of plutonium than conventional reactors, they also increase the risk of nuclear proliferation to a higher level (Jacobson and Delucchi 2011). The world's largest fast breeder reactor, the French Superphoenix, was launched in 1974, connected to the grid in 1986, and closed in 1998 after many technical problems. It only operated for the full power equivalent of ten months and cost €12 billion, excluding decommissioning. No commercial fast breeder reactors are currently producing electricity (Diesendorf 2014).

- *Proliferation of nuclear weapons* is a major problem. Even a regional nuclear war, such as between India and Pakistan, could bring on a 'nuclear winter' resulting in global agricultural collapse for several years and mass starvation (Robock and Toon 2010). There is ample evidence that so-called 'peaceful' nuclear power has contributed, and still contributes, to the proliferation of nuclear weapons. It thus increases the risk of future nuclear war (Diesendorf 2014).
- Nuclear power stations produce *highly radioactive waste* that must be separated from the environment for at least 100,000 years. At present there are no permanent repositories for high-level nuclear waste, although two underground repositories are under construction in Sweden and Finland.
- Nuclear power stations cannot blow up like a bomb, but *can* melt down like Chernobyl and Fukushima and release vast amounts of dangerous radioactivity. There have been 99 nuclear accidents from 1952 to early 2010, totalling $20.5 billion in damages worldwide (Sovacool 2011).
- When nuclear power stations reach the end of their design life, they are so radioactive that dismantling them is a problem. Demolition waste must be stored for thousands of years (Wald 2009). Costs of decommissioning could be as high as the original capital cost of building the nuclear power station (Diesendorf 2014).

These considerations suggest that nuclear fission reactors are unlikely to play a major role in replacing fossil fuels and reducing total carbon emissions by 2050 (Jacobson and Delucchi 2011). Given the above, why would we rush into a nuclear future? The German Advisory Council on Global Change concludes 'the "fossil-nuclear metabolism" of industrial society has no future. The longer we cling to it, the higher the prices will be for future generations' (WBGU 2011, p. 25). Nuclear power is a false solution. It offers little, it offers late, and in the near future (when high-grade uranium ore would be used up), it doesn't offer a big cut in carbon footprint. It is very expensive, and provides its own major risks to future generations. Spending valuable financial resources on nuclear power would be an expensive diversion from fast, effective and less expensive technologies, such as energy efficiency, other forms of demand reduction and renewable energy (Diesendorf 2014). We should not reject denial about the environmental crisis overall, but then deny the serious problems around nuclear power.

Another inappropriate technology to discuss briefly is *geo-engineering*. This is aimed to control climate change without reducing the cause: greenhouse gas emissions. There are two types: solar radiation management (SRM) and carbon dioxide removal (CDR) (Nicholson 2013). The former has received most discussion and is seen as most practical. The observed action of volcanic sulphate aerosols in dropping world temperature has led to proposals for geo-engineering solutions to climate change. Geo-engineering using sulphur means that rather than solving the problem at the source, instead it would pump large amounts of sulphur (as sulphate) into the stratosphere to shade the Earth. However, to compensate for a doubling of CO_2, the required continuous stratospheric sulphate loading would be 5.3 million tonnes per year (Crutzen 2006). World annual production according to the US Geological Survey is 68 million tonnes and world reserves are five billion tonnes (USGS 2010). In terms of resources, a sulphate injection strategy could thus be possible for some years. It has been estimated to cost around $125 billion a year, and would have to be done every one to two years, as the sulphur is naturally removed back into the lower atmosphere (Crutzen 2006). This removal to the lower atmosphere means the sulphur would come back into the Earth's ecosystems and may cause significant acidification of rivers and lakes, though Kravitz et al. (2009) argue this may still be too small to affect most ecosystems.

Apart from the acidity question, this sulphur injection would turn our skies white, would not stop the acidification of the oceans due to increasing CO_2 (which threatens marine food chains), and may affect the Indian monsoon, on which a large part of the world's population depends for food production (Pittock 2009). SRM using sulphur at best can reduce the 'planet's fever' for a certain time, which might allow us a period of grace to reduce emissions. However, many of its proponents seem to assume geo-engineering would be used to continue 'business-as-usual', or even to escalate carbon emissions. To conclude, geo-engineering poses its own major risks, could go catastrophically wrong, and is probably illegal under international law (Nicholson 2013). Geo-engineering is thus part of denial, it denies we need to make hard choices, and pretends we can use a 'magic bullet' to continue business-as-usual (Hamilton 2013). Geo-engineering has no place in any meaningful sustainability.

A final inappropriate (but much discussed) technology to mention is *Carbon Capture and Storage* (CCS). I discuss this briefly here, though in Washington and Cook (2011) we cover it in more detail. CCS is a 'buzz' word, and the reason it is supported by governments is that if it worked it would allow 'business-as-usual' to continue. It too is actually part of denial. Belief in CCS is deluding oneself that we don't need to change, that we can just put the nasty CO_2 somewhere 'away', where it won't do bad things. CCS is not the answer because: (1) It would come too late; (2) It requires huge amounts of energy to liquefy CO_2 and pump it underground; (3) CCS (like nuclear waste) is never disposed of 'forever' (it can escape); (4) the volume of liquefied CO_2 to be disposed of is huge, and would be like an 'oil industry in reverse' in terms of scale (Goodell 2008); (5) It isn't cheap and probably would never be economic without a carbon price of $30 a tonne (Elliston et al. 2014). Recent estimates for both capital and operating costs of coal power with post-combustion CCS systems are approximately *double* those of conventional coal-fired power costs (Diesendorf 2014); (6) Only some areas have a geology that is suitable for CCS. Furthermore, in Australia, coal-fired power stations with CCS could only compete economically with 100 per cent renewable electricity under very uncommon circumstances (Diesendorf 2014).

However, the major reason it is inappropriate is that it *probably won't work* in terms of removing enough CO_2. By 2050 probably only 20–40 per cent of coal-fired CO_2 emissions would be technically suitable for capture. Of these, 5–15 per cent of CO_2 is not captured by the process anyway. If there was 0.5 per cent leakage from these storages (quite likely), then by 2100 there could still be 10 Gt of carbon being released, more than the present rate of carbon emissions (Pittock 2009). We would thus have undertaken a hugely expensive process that did not reduce carbon emissions in a major way. This indeed would be highly inappropriate.

Conclusion

It is no exaggeration to say that whether we reach sustainability will depend on how we deal with technology. Like all tools, technology can be used for good or ill. If we step away from techno-centrism, from modernism and resourcism, from the 'mastery of Nature' ideology, then we can intelligently use the best forms of technology to aid us to reach sustainability. Energy efficiency, energy conservation, renewable energy, technologies to dematerialise and support eco-design and eco-effectiveness are 'appropriate', while nuclear power, geo-engineering and CCS are not. All technology is not evil, but the rampant use of technology without an 'Earth ethics' that respects people and Nature has led us to the brink of disaster. Just as a fire can be your friend and warm your home and cook your food, it can also burn down your town – or even the whole world. 'Sustainability' means using appropriate technologies to live respectfully with the rest of life on Earth, within the ecological limits of the Earth. That means we should encourage helpful technologies, and *abandon* dangerous and inappropriate ones. Unconstrained and rampant use of technology helped create the environmental crisis. However, thoughtful, appropriate and sustainable technology can help get us out of our predicament, and lead us to a sustainable future. To demystify sustainability, we need to accept that technology must be guided by an ecocentric worldview and an Earth ethics. Then it can aid sustainability and be truly appropriate.

References

Assadourian, E. (2010) 'The rise and fall of consumer cultures', in *State of the World 2010: Transforming Cultures from Consumerism to Sustainability*, eds. L. Starke and L. Mastny, London: Earthscan.

AWEA (2012) 'Windpower and reliability: the roles of baseload and variable resources: American Wind Energy Association', online, available at: www.awea.org/learnabout/publications/upload/Baseload_Factsheet.pdf (accessed 11 February 2012).

BNEF (2013) 'Clean energy investment – Q2 2013 fact pack', London: Bloomberg New Energy Finance, online, available at: http://about.bnef.com/presentations/clean-energy-investment-q2-2013-fact-pack/ (accessed 2 December 2014).

Boyle, G. (ed.) (2012) *Renewable Energy: Power for a Sustainable Future* (third edition), Oxford: Oxford University Press.

Brown, L. (2006) *Plan B 2.0: Rescuing a Planet under Stress and a Civilization in Trouble*, New York: W. W. Norton and Company.

Crutzen, P. (2006) 'Albedo enhancement by stratospheric sulfur injections: a contribution to resolve a policy dilemma', *Climatic Change*, vol. 77, nos 3/4, pp. 211–220.

Daly, H. (1991) *Steady State Economics*, Washington: Island Press.

Delucchi, M. and Jacobson, M. (2011) 'Providing all global energy with wind, water, and solar power, part II: reliability, system and transmission costs, and policies', *Energy Policy*, vol. 39, pp. 1170–1190.

Diesendorf, M. (2007) *Greenhouse Solutions with Sustainable Energy*, Sydney: UNSW Press.

Diesendorf, M. (2011) 'A cheaper path to 100% renewables', *Climate Spectator*, 21 October, 2011, online, available at: www.climatespectator.com.au/commentary/cheaper-path-100-renewables (accessed 22 July 2014).

Diesendorf, M. (2013a) 'Baseload power is a myth: even intermittent renewables will work', *Conversation*, 10 April, online, available at: https://theconversation.com/baseload-power-is-a-myth-even-intermittent-renewables-will-work-13210 (accessed 26 September 2013)

Diesendorf, M. (2013b) 'The economics of nuclear energy', in *Nuclear Power and Energy Security in Asia*, eds. R. Basrur and C. Koh, London: Routledge.

Diesendorf, M. (2014) *Sustainable Energy Solutions for Climate Change*, Sydney and Abingdon, UK: NewSouth Publishing and Routledge-Earthscan.

Elliston B., Diesendorf M., MacGill I. (2012) 'Simulations of scenarios with 100% renewable electricity in the Australian national electricity market', *Energy Policy*, vol. 45, pp. 606–613.

Elliston, B., MacGill, I. and Diesendorf, M. (2014) 'Comparing least cost scenarios for 100% renewable electricity with low emission fossil fuel scenarios in the Australian national electricity market', *Renewable Energy*, vol. 66, pp. 196–204.

EPIA (2013) 'Global market outlook for photovoltaics 2013–2017, European Photovoltaic Industry Association', online, available at: www.epia.org/news/publications/ (accessed 3 October 2013).

GEA (2012) *Global Energy Assessment – Toward a Sustainable Future*, Cambridge, UK: Cambridge University Press, online, available at: www.globalenergyassessment.org (accessed 16 September 2013).

Geodynamics (2011) 'Geodynamics power from the Earth annual report 2010–2011', online, available at: www.geodynamics.com.au/IRM/content/PDF/Geodynamics_AnnualReport_2011_Web.pdf (accessed 22 November 2011).

Goodell, J. (2008) 'Coal's new technology: panacea or risky gamble', *Yale Environment 360* (14 July 2008), online, available at: http://e360.yale.edu/feature/coals_new_technology_panacea_or_risky_gamble/2036/ (accessed 22 July 2014).

GWEC (2010) *Global Wind Report: Annual Market Update 2010*, Global Wind Energy Council, online, available at: http://gwec.net/wp-content/uploads/2012/06/GWEC_annual_market_update_2010_-_2nd_edition_April_2011.pdf (accessed 2 December 2014).

Hamilton, C. (2013) *Earthmasters: The Dawn of the Age of Climate Engineering*, New Haven: Yale University Press.

IBI (2012) 'How much carbon can biochar remove from the atmosphere?', International Biochar Initiative, online, available at: www.biochar-international.org/biochar/carbon (accessed 22 July 2014).

IEA (2010) 'Technology roadmap: Concentrating solar power', International Energy Agency, online, available at: www.iea.org/publications/freepublications/publication/csp_roadmap_foldout.pdf (accessed 22 July 2014).

IEA (2012) *World Energy Outlook 2012*, Paris: OECD/International Energy Agency.
IR (1996) 'Environmental impacts of large dams: African examples', International Rivers website, online, available at: www.internationalrivers.org/africa/environmental-impacts-large-dams-african-examples (accessed 22 July 2014).
IRENA (2012) 'Solar photovoltaics, renewable energy technologies: cost analysis series', Vol. 1: Power Sector, Issue 4/5, International Renewable Energy Agency, online, available at: www.irena.org/DocumentDownloads/Publications/RE_Technologies_Cost_Analysis-SOLAR_PV.pdf (accessed 3 October 2013).
Jacobson, M. and Delucchi, M. (2011) 'Providing all global energy with wind, water, and solar power, part i: technologies, energy resources, quantities and areas of infrastructure, and materials', *Energy Policy* vol. 39, pp. 1154–1169.
Kanter, J. (2009) 'European solar power from African deserts?', *New York Times*, 18 June.
Kemp, M. and Wexler, J. (eds) (2010) 'Zero carbon Britain 2030: a new energy strategy'. Technical Report, Centre for Alternative Technology, online, available at: http://zerocarbonbritain.com (accessed 22 July 2014).
Klein, J. (2010) 'Comparative costs of California central station electricity generation', California Energy Commission, online, available at: www.energy.ca.gov/2009publications/CEC-200–2009–017/CEC-200–2009–017-SF.PDF (accessed 26 September 2013).
Kravitz, B., Robock, A., Oman, L., Stenchikov, G. and Marquaqrdt, A. (2009) 'Sulfuric acid deposition from stratospheric geoengineering with sulfate aerosols', *Journal of Geophysic Research*, vol. 114, no. D14.
Kumar, A. and Ramana, M. (2009) 'The safety inadequacies of India's fast breeder reactor', *Bulletin of the Atomic Scientists*, 21 July.
Lehmann, J., Gaunt, J. and Rondon, M. (2006) 'Bio-char sequestration in terrestrial ecosystems – a review', *Mitigation and Adaptation Strategies for Global Change*, vol. 11, pp. 403–427.
Lovins, A. (2006) Quoted in WADE (2006) *World Survey of Decentralised Energy 2006*, Washington, DC: World Alliance for Decentralised Energy, online, available at: http://localpower.org/nar_publications.html (accessed 22 July 2014).
Lovins, A. and Sheikh, I. (2008) *The Nuclear Illusion*, Colorado: Rocky Mountain Institute, online, available at: www.rmi.org/rmi/Library/E08–01_NuclearIllusion (accessed 22 July 2014).
Makhijani, S. and Ochs, A. (2013) 'Renewable energy's natural resource impacts' in *State of the World 2013: Is Sustainability Still Possible?*, ed. L. Starke, Washington: Island Press.
Meadows, D., Randers, J. and Meadows, D. (2004) *Limits to Growth: the 30-year Update*, Vermont: Chelsea Green.
Mudd, G. and Diesendorf, M. (2008) 'Sustainability of uranium mining and milling: toward quantifying resources and eco-efficiency', *Environmental Science and Technology*, vol. 42, no. 7, pp. 2624–2630.
Murphy, T. (2013) 'Beyond fossil fuels: assessing energy alternatives', in *State of the World 2013: Is Sustainability Still Possible?*, ed. L. Starke, Washington: Island Press.
Nicholson, S. (2013) 'The promises and perils of geoengineering', in *State of the World 2013: Is Sustainability Still Possible?*, ed. L. Starke, Washington: Island Press.
NRC (2007) *Environmental Impacts of Wind-Energy Projects*, Washington, DC: National Research Council of the US National Academy of Science, online, available at: https://download.nap.edu/catalog.php?record_id=11935 (accessed 5 February 2012).
Pittock, A. B. (2009) *Climate Change: the Science, Impacts and Solutions*, Collingwood, Vic/London: CSIRO Publishing/Earthscan.
REN21 (2013) *Renewables 2013 Global Status Report*, Renewable Energy Policy Network

for the 21st Century, online, available at: www.ren21.net/REN21Activities/GlobalStatusReport.aspx (accessed 3 October 2013).

Riedy, C. and Diesendorf, M. (2003) 'Financial subsidies to the Australian fossil fuel Industry', *Energy Policy*, vol. 31, pp. 125–137.

Robock, A. and Toon, B. (2010) 'South Asian threat? Local nuclear war = global suffering', *Scientific American*, vol. 302, no. 1, pp. 74–81.

Rolston III, H. (2012) *A New Environmental Ethics: The Next Millennium of Life on Earth*, London: Routledge.

Schumacher, E. F. (1973) *Small is Beautiful: A Study of Economics as if People Mattered*, New York: Blond and Briggs.

Siemens (2010) 'Ultra HVDC transmission system', online, available at: www.energy.siemens.com/co/en/power-transmission/hvdc/hvdc-ultra/#content=Benefits (accessed 22 July 2014).

Sovacool, B. (2011) *Contesting the Future of Nuclear Power: a Critical Global Assessment of Atomic Energy*, Singapore: World Scientific Publishing Co., online, available at: www.worldscientific.com/worldscibooks/10.1142/7895 (accessed 22 July 2014).

Taylor, P. (2010) *The Biochar Revolution: Transforming Agriculture and Environment*, Mt Evelyn, Australia: Global Publishing Group.

USEIA (2011) *Annual Energy Review*, US Energy Information Administration, Washington, DC: USEIA.

USGS (2011) 'Sulphur', online, available at: http://minerals.usgs.gov/minerals/pubs/commodity/sulfur/mcs-2011-sulfu.pdf (accessed 6 March 2012).

Wald, M. (2009) 'Dismantling nuclear reactors', *Scientific American*, 26 January.

Washington, H. and Cook, J. (2011) *Climate Change Denial: Heads in the Sand*, London: Earthscan.

WBGU – German Advisory Council on Global Change (2011) 'World in transition: a social contract for sustainability'. Summary for Policy-Makers. Berlin: WBGU, online, available at: www.wbgu.de/fileadmin/templates/dateien/veroeffentlichungen/hauptgutachten/jg2011/wbgu_jg2011_en.pdf (accessed 22 July 2014).

Weizsäcker, E. von, Hargroves, K., Smith, M., Desha, C. and Stasinopoulos, P. (2009) *Factor 5: Transforming the Global Economy through 80% Increase in Resource Productivity*, London Earthscan.

Woolf, D., Amonette, J., Street-Perrott, F., Lehmann, J. and Joseph, S. (2010) 'Sustainable biochar to mitigate global climate change', *Nature Communications*, vol. 1, no. 5, pp. 1–9.

Wright, M. and Hearps, P. (2010) 'Australian sustainable energy stationary energy plan', Melbourne: University of Melbourne Energy Research Institute/Beyond Zero Emissions, online, available at: http://media.bze.org.au/ZCA2020_Stationary_Energy_Report_v1.pdf (accessed 2 December 2014).

WWF (2011) 'The energy report: 100% renewable energy by 2050', Switzerland: WWF, online, available at: http://wwf.panda.org/what_we_do/footprint/climate_carbon_energy/energy_solutions/renewable_energy/sustainable_energy_report/ (accessed 3 February 2012).

11
SOLUTIONS FOR SUSTAINABILITY

> *Just be present, when you're worrying about whether you're hopeful or hopeless or pessimistic or optimistic, who cares? The main thing is that you are showing up, that you're here, that you're finding ever more capacity to love this world, because it will not be healed without that.*
>
> *(Joanna Macy 2012)*

Can we know what future generations will want?

Can we know what future generations will want? This question has often been used as an excuse for not solving the environmental crisis. For example, Solow (1993, p. 180) argues 'the tastes, the preferences, of future generations are something we don't know about'. However, Schmidheiny (1992, p. 11) responds:

> Some argue that we have no responsibility for the future, as we cannot know its needs. This is partly true. But it takes no great leap of reason to assume that our offspring will require breathable air, drinkable water, productive soils and oceans, a predictable climate, and abundant plant and animal species on the planet they will share.

I support this conclusion. Future generations, like our own, will be *dependent on Nature*, they will need clean air, soil and water. They will need functioning ecosystems, they will need robust functioning nutrient cycles, and they will need pollinators to pollinate their crops. They will need mangroves and salt marshes for fish nurseries and to protect coastal areas from storms. They will also need wild places to recharge their souls, inspire creativity, and to counter 'Nature deficit disorder' (Louv 2005). They too I believe will also want a society that addresses humanity's underlying desire for peace, justice and equity. These needs are very unlikely to change. Physical needs will be as essential for future generations as they are today. Our psychological and spiritual needs have been with humanity since we evolved, hence I am sure they will continue in our descendants. We thus *can* have a pretty good idea what future generations will both need and want. Understanding this, we are currently failing in our custodianship on all counts. Sustainability means dramatically improving that custodianship.

Solutions – pluralism vs specificity

This chapter is about 'solutions for sustainability'. However, there are different views on how to undertake solutions. One approach is about *pluralism* and comes from a postmodern perspective. This argues that there are many different visions, approaches and pathways that we should be acting on for sustainability. All are often portrayed as 'equally valid'. One of my reviewers for this book argued this, stating: 'the sustainability literature of the last 10–15 years has emphasised the different paths to sustainability depending on the physical and cultural situation and priorities of the particular groups'. Indeed, the suggestion of a single vision of (or path to) sustainability, or that there are *specific solutions*, was described by this reviewer as a 'domineering tendency'. On the face of it, the claim that there are many different paths and visions sounds great. It means that everyone's approach is correct, so you don't have to push for any particular key strategies.

Others disagree, they think we should be advocating the *best* vision and the most practical solutions to solve the problems. Hill (1992) warns that we should beware 'deceptive simplicity' and seek 'profound simplicity', with 'solutions that often involve ways to avoid problems or that turn them into benefits'. Jickling (2009) argues against the dominant view in post-structuralism within academia that education 'has no meaning and no ends' other than those subjectively ascribed to it. Wals (2010, p. 150) notes there is a conflict between a deep concern and urgency about the state of the planet, and a conviction that it is 'wrong to persuade, influence or even educate people towards pre- and expert-determined ways of thinking and acting'. Rickinson (2003) and Chawla and Cushing (2007) call for 'clear objectives and ends' in order to make education for sustainable development effective in addressing environmental problems. Kopnina (2012) notes that:

> research on education for sustainable development is dominated by calls for pluralistic, emancipatory or transactional forms of education that encourage co-creation of knowledge ... and encourage multiple perspectives and critical dialogue on the very concept of sustainable development and education for sustainable development (Gough and Scott 2007; Wals 2007).

Kopnina concludes that the pluralistic approach to education for sustainable development 'signals scholarly departure from the "real world" dilemmas concerned with environmental degradation'. She observes that 'academic relativism about education for sustainable development might in fact be undermining the efforts of educating citizens in the importance of valuing and protecting the environment'. She notes that scholars of sustainable development seem increasingly engaged in academic debates about the 'dangers of dogmatic thinking'. She wonders if such theorists might be failing to: 'see the (still standing) forest behind the (receding) trees'.

I share that deep concern. I unashamedly argue here for a framework of *specific solutions*. In regard to sustainability, for almost three decades society has fiddled while Rome burned around us. It seems also that many in academia have been playing along in the string section. The dominance of postmodernism within academia has led to arguments for pluralistic approaches, and to a belief that any over-arching 'grand narrative' was wrong and should be opposed. I believe this is why so much of the debate around 'sustainability' and 'sustainable development' has become tokenism, why Gare (1995) noted that postmodernism was good at analysing the problems of modernism, but powerless to provide solutions. As was discussed in Chapter 8, we *do* in fact need a grand narrative of sustainability now if we are to solve the environmental crisis. And that grand narrative cannot remain one of endless growth (as was that of 'Our Common Future', WCED 1987). It cannot remain one of refusing to accept ecological limits and physical laws. It cannot remain one of ignoring the centrality of worldview and ethics.

Hence I believe there *is* a single vision of sustainability, which is *accepting reality*, accepting that society is now hugely *un*sustainable, and undertaking the 'Great Work' of moving to a truly sustainable (ecologically, socially and economically in the long-term) future. Under that vision there are of course many goals, strategies and parts to the solutions. My conclusion (echoed by others such as Kopnina 2012) will clearly be unpalatable to some in academia. However, I write here for the educated layperson and seek to demystify sustainability. I seek to provide real solutions to real problems, rather than theorise (under the current latest theory), procrastinate, or deny. Each reader must decide *for themselves* whether the solutions listed are too 'domineering' or actually a useful way forward. I believe there are clear solutions to reach a sustainable future, and list the key solution frameworks. The first part of a way forward is to consider what 'sustainability' – if it is to be meaningful – *cannot be*.

Demystification – what meaningful sustainability *cannot* be

If 'sustainability' is to be meaningful in terms of solving the environmental crisis and leading us to a long-term future in harmony with Nature, then there are some things 'sustainability' cannot be. Below I examine what it cannot be:

1   It cannot be sustaina*babble* (Engelman 2013) it cannot mean all things to all people. It cannot remain undefined and delightfully vague. Vagueness is not a 'strength' but a flaw that has let sustainability slip into tokenism.
2   It cannot be a *denial of reality*, it has to be about 'realism'. That means we accept the problems we have, and seek to solve them. Hence sustainability cannot be

about denial, the two are mutually exclusive. Sustainability also should not be about esoteric theory and modelling, especially when these are based on ideology, such as modernism (and sometimes postmodernism) and the 'mastery of Nature' worldview. Theory that does not assist sustainability, or that twists ecological reality and denies the environmental crisis (see Chapter 5), needs to be modified or abandoned.

3   It cannot ignore the *ecological limits of the Earth*, if it does then it cannot actually be 'sustainable'.

4   Accordingly, it *cannot be about endless physical growth* on a finite planet. It cannot be about 'more', it has to be about 'enough'. Any supposedly 'sustainable' project that is based on continuing physical growth is clearly a delusion and 'greenwash'. It cannot be sustainable in a world that has far exceeded its limits. That includes a 'green economy' that is still a physical growth economy (see Chapter 4). Our non-physical culture, our wisdom, our spirituality *can* keep growing, but our physical impact cannot. Denial of this just leads us closer to collapse, as other cultures discovered. We cannot grow our way physically into sustainability, for growth is the root cause of our *un*sustainability. Yet this is the underlying assumption to 'Our Common Future' (WCED 1987) and many government sustainability strategies. Today, 'more' is not better, but far worse.

5   Hence, if we mean *growth* by 'development' (and most people do), then sustainability cannot be the same as 'sustainable development'. 'Sustainable growth' on a finite planet is an oxymoron, so if that is the meaning you attach to 'development', then 'sustainable development' also becomes a contradiction in terms. Further growth quantitatively is not sustainable, while growth qualitatively is. In other words, growth in our culture, art, and wisdom can continue, and hopefully will do so. Hence my focus in writing this book has been on *sustainability* and not on 'sustainable development', the common meaning of which has come to be seen as growth. It is I believe a mistake to equate them.

6   Sustainability *cannot be about a 'weak sustainability'* that believes we can substitute money for ecosystem services. This breaks fundamental ecological reality. At a minimum it has to be about 'strong sustainability' that retains ecosystem services and natural capital for humanity. However, our solutions should go beyond that and be for *all* of Nature, hence it should be 'strongest sustainability' (see Chapter 3), which accepts the intrinsic value of Nature and believes in Earth jurisprudence and eco-justice (see Chapter 8).

7   Sustainability *cannot be ethics-free*. It cannot be based on an anthropocentric 'human supremacy' approach, where humanity always seeks to be the 'master'.

So there are at least seven things that sustainability – to be meaningful – cannot be. What then 'should' it be?

What sustainability *should* be?

So after ten chapters seeking to demystify sustainability, we return to this question. We all stand at a point of decision. What is sustainability? We must now demystify and define it. If we don't, it becomes tokenism and virtually meaningless. 'Sustainability' – first and foremost – must be about *solving the environmental crisis*, which is now rapidly worsening. To do this, sustainability must be about 'realism', a realism founded on what mainstream environmental science is telling us, that we are degrading the life support systems that maintain our civilisation. It has to be a realism founded on an understanding that our species, *Homo sapiens*, has a key failing: denial. If we acknowledge our failing, then we can correct our tendency to believe unsustainable delusions. We have seen that our economy is broken, our society is broken, and that the ecosystems that support us are breaking. We have seen too that the worldview and ethics of Western society are key causes of *un*sustainability. In ecological terms, we are vastly over-populated and over-consuming. Also, sadly we have seen that mainstream society denies all of these. We delude ourselves that 'everything's fine'. Yet the interest in 'sustainability' within society today shows that many of us (at least in our inner hearts) realise that everything is not truly fine. So sustainability is the task of *healing these broken things*. This is what a meaningful sustainability should be. It has to be about creating a culture that lives in harmony with Nature (and each other) into the future. We have to break the denial dam, and let the solutions flow free. 'Sustainability' needs to be the vision that inspires us to take difficult decisions. It needs an 'Earth ethics' (Chapter 8) based on an ecocentric worldview, where we respect both each other and the Nature of which we are a part, and to which we feel respect and responsibility.

The environmental crisis is solvable, we *can* reach sustainability. Not by pretending and deluding ourselves. Not by saying it's 'too hard'. Not by prostrating ourselves before a neoliberal 'Free Market' god and hoping it will save us. Rather, it will be by accepting the need, and rising to the challenge: enacting solutions for sustainability. Meadows *et al.* (2004) invite us to imagine how different things would be if we accepted that a global transition to a sustainable society was possible. Making broader change seems to involve three things:

1   a big idea of how things could be better;
2   a commitment to move beyond individual actions and join with others to make it real;
3   then *real* action must follow (Leonard 2013, p. 250).

The missing ingredient is collective engagement for political and structural change. However, if we move past denial and accept our problems, then we can commence the 'Great Work' (Berry 1999) of being sustainable. Sustainability is thus healing the planet, along with our society and our maladjusted economy. This may take time. We should accept that true sustainability may not arrive for decades or even centuries (Engelman 2013, p. 14). However, the next few decades will be crucial. The future we reach together will essentially be set in motion *now*.

Is it too late? Optimism, pessimism and *realism*

We have to *accept reality* to solve our problems. Vitousek et al. (1997) noted: 'We are the first generation with tools to understand the changes in the Earth's systems caused by human activity and the last with the opportunity to influence the course of many of the changes now rapidly under way'. It is time to use those tools to make that change. It is time to 'grow up' and drop our delusions of being 'masters of Nature'. We never 'conquered' Nature, we just raided fossil fuels to cause the environmental crisis (Greer 2009). We stole fossil energy from the past, and (due to the consequent impacts) we are now robbing the future, making it unsustainable for those who follow us. Sustainability demands we reassess ourselves. Environmentalism has now become the 'most significant human, moral and theological problem confronting the contemporary world' (Collins 2010). If we don't face up to the ecological crisis 'we will have no future as sane, ethical and spiritual beings' (Swimme and Berry 1992). It is time to stop being a 'seriously dumb species' (Soskolne 2008) if we are going to reach meaningful sustainability.

So, how do we do this? How to solve the environmental crisis and reach sustainability? First, it is essential to say it is *not hopeless*! There is always hope, provided it is tied in to facing up to reality. Second, it will not be *easy*, as we need to directly challenge the dominant worldview, ethics, power structures, economic system, and cultural and religious underpinnings of our society. However, we all love a challenge! It would have been far easier 30 years ago when William Catton (1982) wrote the visionary 'Overshoot'. Yet we fell into denial, and failed to take real action. No generation has faced a challenge of the complexity, scale and urgency we now face (Brown 2011). Now we must act quickly. There can be no doubt of the urgency of solving the environmental crisis. As Chapter 5 and other chapters have shown, we really *do* have a problem, a huge one. Whether we solve this or not will be played out this century, almost certainly before mid-century (Meadows et al. 2004). It won't be over in a moment, and in fact it is likely to be the 'long emergency' (Orr 2013, p. 279). It has also been called the 'long descent' (Greer 2008), a form of decline or collapse (as our energy and material use must decrease). This descent may proceed in stages, and hopefully we can *learn about sustainability* at each step in the descent. This means we have to accept that we cannot continue 'business-as-usual'. We now need to take a serious look at ourselves, our worldview and ethics, our ideologies, our societies, our economy, our technologies, and our denial. The Millennium Ecosystem Assessment (MEA 2005) noted that significant changes to policies, institutions and practices can mitigate many of our impacts. However, they also noted that the changes required are large and *not currently underway*.

Humans are intelligent (often!), and when committed can (and have) solved huge problems. The Second World War, the Marshall Plan to rebuild Europe after that war, and the Space Race are examples of successful major actions that some would have said were impossible. Atlee (2003) describes 'co-intelligence' as the human capacity to generate creative responses, and it is this we must foster. We face a huge problem, but it is also a major 'opportunity' to build a truly sustainable future. We should be embracing these solutions for many reasons. For example, we need to stabilise population for reasons of food security, biodiversity protection and water supply, but this will also reduce carbon emissions and improve quality of life and reduce impacts on the poor. By acting responsibly on climate change, we can solve the energy crisis, and also take a big step towards slowing the extinction crisis. We should stop burning fossil fuels for health reasons as well as climate (Haswell Elkins and Washington 2014).

Hulme (2009) refers to 'wicked' problems, which have no single simple solution. Rather than just one 'silver bullet' to solve the problem, we need multiple solutions or 'silver buckshot'. However, to continue the comparison, while there is no 'magic bullet', the question is where do you aim the shot, and how fine do you make it? I don't accept the pluralist argument that all solutions are equally valid, or that it's fine if some contradict others. This promotes dithering. I list here *nine key solutions* that I believe underpin the 'Great Work' of sustainability. There is good agreement on these by environmental scientists and scholars over several decades. Many books have been written about our predicament. Some notable ones are Catton (1982) 'Overshoot', Ehrlich and Ehrlich (1991) 'Healing the Planet', Berry (1999) 'The Great Work', Soskolne (2008) 'Sustaining Life on Earth' and Brown (2011) 'World on the Edge'. It is notable how much agreement there is in such books as to both the problems and the solutions. It's not like the solutions have not been pointed out, it is that they have *not been acted on*. Edwards (2005) summarises many of the policies and principles that have been developed on sustainability in many countries. These include the Ontario ORTEE principles; the excellent Netherlands National Environmental Policy Plan; the Natural Step, the Houston Principles; the CERES Principles; the Forestry Stewardship Council Principles; the Asilomar Declaration for Sustainable Agriculture; the Hannover Principles; Todd's Principles of Ecological Design; the Sanborn Principles; the LEEDing Edge; Deep Ecology Principles; Permaculture Principles; and of course the inspirational 'Earth Charter'.

Edwards (2005) points out there are remarkable similarities in the key values expressed in these principles. He concludes there are seven common themes: *stewardship* (an ecological ethic); *respect for limits* (living within Nature's means and preventing waste and pollution); *interdependence* (connectivity within Nature, on which humanity depends); *economic restructuring* (that safeguards ecosystems, though rarely depicted sadly as going as far as a steady state economy); *fair distribution* (equality); *intergenerational perspective* (taking a long term view); and *Nature as a model and teacher* (which includes respecting the 'Rights of Nature'). These themes are indeed crucial as solutions for sustainability. If we just acted on these themes alone, it would be a vast step forward. However, there are other things we must also do, as we shall see.

Robinson (2013, pp. 374–375) makes the point:

> if we were to do everything right, starting this year and continuing for the next several decades ... decarbonise, and to conserve, restore, protect, replace

and so on – then we could do it ... it is physically possible ... are we going to do everything right in the rest of the twenty-first century?... it looks very unlikely. We just aren't that good, either as a species or a civilization.... We are going to do damage in the twenty-first century, possibly big damage.

Indeed we are already doing damage. It might not involve a dramatic fall, so much as a long decline (Greer 2008), a long emergency (Orr 2013). Or it may come with a crunch, the 'perfect storm' caused by collision of changing climate; spreading ecological disorder (including soil loss, deforestation, water shortages, species loss, ocean acidification); population growth; unfair distribution of the costs, risks and benefits of economic growth; ethnic and religious tensions; and proliferation of nuclear weapons (Orr 2013, p. 279). So the question should be: how much damage will we let happen? And 'How much will we save?'. How much of the biosphere will be saved, that is the real question (Robinson 2013, p. 375). Can we prevail over our destructive private interests, when the common good of humanity and the biosphere is at stake? Ehrlich and Ehrlich (2013) conclude we can avoid collapse – provided we take the needed actions to deal with long-term threats. Will we? It remains an open question (Robinson 2013, p. 377). In the end it is up to *us*. There is nobody else.

Neither 'fear of pessimism' (doom and gloom), nor 'dogged determination' to remain optimistic, are reasons to understate our predicament. I have sat through many talks where the speaker insists how 'optimistic' he or she is, and books likewise (e.g. Edwards 2010). However, Ehrenfeld (1978) notes that optimism is necessary for those who 'attempt the impossible'. Optimism and pessimism are equal distractions from what we need: *realism*, a commitment to Nature and to each other, and a determination not to waste more time (Engelman 2013, p. 16). 'Feeling you have to maintain hope can wear you out' notes Joanna Macy (2012), who advises:

> Just be present, when you're worrying about whether you're hopeful or hopeless or pessimistic or optimistic, who cares? The main thing is that you are showing up, that you're here, that you're finding ever more capacity to love this world, because it will not be healed without that.

Kathleen Moore and Michael Nelson (2013, p. 229) beautifully express what we need now – 'moral integrity':

> But to think that hope and despair are the only two options is a false dichotomy. Between them is a vast fertile middle ground, which is integrity; a matching between what we believe and what we do. To act justly because we believe in justice. To act lovingly toward children because we love them. To refuse to allow corporations to make us into instruments of destruction because we believe it is wrong to wreck the world. This is moral integrity. This is a fundamental moral obligation – to act in ways that are consistent

with our beliefs about what is right. And this is a fundamental moral challenge – to make our lives into works of art that embody our deepest values.

So is it too late? **No, *it's never too late***. Any action we take is better than none. Foresters say the best time to plant a tree is decades ago, the second best time is today (Zencey 2013, p. 83). Just as water can erode a rock, the continuing pursuit of cultural change can add up to much more than its parts. Seeds sown today may take root when humanity reaches for solutions to rebuild, as systems unravel under the unbearable burden of sustaining a global consumer economy (Assadourian 2013, p. 124). Ecologist Paul Ehrlich notes that there are thresholds in human behaviour when cultural evolution moves rapidly, so that: 'When the time is ripe, society can be transformed virtually overnight' (in Daily and Ellison 2002).

So – we should all accept the invitation to abandon pessimism and depression, which really means one will do nothing to make things better. Abandon also gung-ho optimism (and denial), which downplays the severity of the mess we have created, and pretends that the solutions are simple. They won't be simple, they will be hard, but they *are* possible. Instead, become a 'realist', accept we have major problems that we must solve to become sustainable. Not in a decade, not in a year. *Now*. It is never too late to do the right thing.

Solutions

The chapters on economic (Chapter 4), ecological (Chapter 5) and social (Chapter 6) sustainability have listed many solutions under what these strands of sustainability should be. Chapter 7 (population and consumerism), Chapter 8 (worldview), Chapter 9 (denial) and Chapter 10 (appropriate technology) have listed others. Clearly, not all of these individual solutions will be repeated here, where I shall focus on key solution frameworks already raised in previous chapters). To be realistic, the most I can attempt here is to cover the broad structure of sustainability solutions. One can easily write books on each (and many have!). Given that people get bogged down in the detail, I will conclude each section with a short 'What you can do!'. Those who do not wish to revisit individual solution frameworks may wish to advance to the section on the 'Great Work'. At the end of the chapter I shall return to the key question of 'Can we demystify sustainability?'.

1 Worldview, ethics, values and ideologies

Chapter 8 showed the Western modernist worldview to be broken, it can even validly be called 'evil', as it is anti-life and fundamentally unsustainable (Collins 2010). We desperately need to change our worldview, ethics, values and ideologies. Humanity is faced with the need for rapid cultural change to avoid a collapse of civilisation (Ehrlich and Ehrlich 2013). This is a complex problem of 'conscious evolution' (Ehrlich and Ornstein 2010) or cultural evolution to create a thoughtful, ecologically-literate human, motivated by a 'sense of responsibility to the planet'

(Soskolne et al. 2008). The change in worldview is unlikely to come from governments and business, the media, or even from the education system. That leaves us – 'we the people'. When enough of us change our worldview, then governments, business and the education system will follow. To quote Gandhi: 'We must become the change we want to see in the world'.

> **What you can do!**
>
> Talk about this! Become a champion for changing society's worldview and ethics. Talk about Nature's 'intrinsic value', environmental ethics and the need for both social justice and eco-justice. Raise it with friends and family. *Argue the need for change.* You may find more allies than you thought?

2 Redesigning ourselves to enable change

It has been pointed out by social ecologists such as Stuart Hill that our species is psychosocially undeveloped, that the path to sustainability will also require personal transformation and development and a cultural or 'psychosocial' evolution within most societies (Hill 1991, 2014). Hill believes this is possible, but we need to recognise such evolution is a key part of the solution. As Chapters 4 and 5 have shown, most of us have grown up within a society that is unsustainable, one based on an idea of endless growth, competition and inequality. Accordingly, most of us have (to greater or lesser extent) been conditioned or damaged within our inner selves. Children tend to be raised (mostly unintentionally) in oppressive ways by parents and others (who were themselves damaged by being raised that way). Hence most of us suffer a deep inner distress. However, for the first time in history we have the means to develop psychosocially, and Hill argues we have the theory and practice to recover from distress, and that this is necessary if we are to reach sustainability (Hill 1991). This is why we need to *redesign ourselves*.

We will have to pay more attention to how we raise and educate our children, to social attitudes towards 'worth' and status, and to the design of institutional structures and processes. Our institutions should support sustainability rather than remain a barrier (Hill 2014, p. 408). Hill (2014) argues that we need to build our 'personal capital' or personal sustainability to enable us to have the inner strength to create real change. This means empowerment, awareness, creative visioning, worldview clarification, developing our responsibility and spontaneity. We need to develop our ability to care, feel empathy, love, be mutualistic, value equity, and to extend our compassion to other species and the planet (Hill 2014). We need to support our personal development and continue a process of life-long learning. This will enable us to build our social capital (see Chapter 4). We will also need to develop our inter-cultural or 'interpersonal capital' to build trust and cooperation. We need to be 'proactive' rather than reactive, and accept the need to fundamentally redesign our society, rather than just 'increase efficiency' and find less dangerous substitutes to solve problems (Hill 2014). Too much of the discussion and action around 'sustainability' has really just continued 'business-as-usual'. We also need to put 'change' in perspective. Most of the changes towards sustainability that we can make individually will realistically be small, but they can still be very *meaningful*. We should celebrate these changes publicly, so they can be copied by others (Hill 2001). Big projects (e.g. giant dams) may be 'newsworthy', but are not necessarily the best action towards sustainability. Real sustainability will come from the net result of a great many of us acting collectively for change. Redesigning ourselves will require many of us to radically rethink our worldview and ethics. It will definitely mean examining our denial. It will mean thinking about our 'spiritual' capital as much as our built and human capital.

> **What you can do!**
>
> Think about how you can 'redesign yourself'! What of your worldview and ethics? What of your empathy and compassion? What about your *own* denial, your own passivity and lack of action? How can you empower yourself to do more? What of how you educate and raise your children? What small (but meaningful) actions can you take to make the world more sustainable?

3 Population

As Chapter 7 showed, we cannot talk about sustainability unless we talk about overpopulation. There are just too many of us living with a high and increasing consumption. Overpopulation *can* in fact be tackled by various strategies (Engelman 2012). If we are going to accept reality, then we need to accept all of it. Population growth is a key contributor to the environmental crisis. We need to stop denying this and develop rational, ethical, humane and non-coercive, ecologically-based population policies. 'More people' is no longer 'better', but far worse. It keeps us from reaching a sustainable future.

> **What you can do!**
>
> Dare to talk about the double whammy of overpopulation and overconsumption on a finite planet, that the two together multiply to cause unsustainable impact. Argue for an ecologically sustainable population target for your nation. Have no more than two children. Talk about how our ecological footprint is at least 1.5 Earths. Explain that we need action on both overpopulation and overconsumption.

4 Consumerism and the growth economy

Is it 'economic' to solve the environmental crisis? This has the question back to front. We should ask whether it is economic *not* to solve it. It may cost a lot of money, but it won't cost more than we already spend on the military (Pittock 2009). Brown (2011) in fact estimates it will cost about an *eighth* of the world's military budget. Arguments of false economic expediency also need to be refuted. Stern (2006) showed that failing to act on climate change would end up costing us far more than acting now. The same applies to other environmental problems. Saving civilisation means restructuring the economy at 'wartime speed'. It also means restructuring taxes to get the market to tell the 'ecological truth' (Brown 2011). Chapter 4 showed our economy was broken and fundamentally unsustainable, but there is an alternative we can move to – a 'steady state economy'. Chapter 7 showed that the consumer culture is fundamentally unsustainable. It does not make us happier or increase our true quality of life. Our throwaway society is literally trashing the Earth. It is overdue for us to come to grips with and reject this created ideology of consumerism. We need to move to 'cradle to cradle' eco-design, reject throwaway products, and we need to rethink the need for 'more stuff', then we should reuse before recycle. Groups such as the No Impact Project (online, available at: www.noimpactproject.org) and 'The Compact' (online, available at: www.sfcompact.blogspot.com) are citizen movements that reject consumerism. The New Economics Foundation (online, available at: www.neweconomics.org/) is another resource (NEF 2012) as is CASSE (online, available at: http://steadystate.org/).

> **What you can do!**
>
> Challenge the mantra of the growth economy and consumerism. Point out there is another way – a *steady state economy*. Argue for tax and subsidy shifts, for a Green GDP, and for minimum and maximum incomes. Argue for a truly 'honest market' that factors in all environmental and social costs. Reuse and recycle products. If possible grow your own food. Argue against rampant consumerism, concentrate on well-being rather than more 'stuff'. Consume less, shop less and live more!

5 Solving climate change

A key task of any meaningful concept of 'sustainability' is dealing with human-caused climate change and its denial (Washington and Cook 2011). Human-caused climate change is a reality we must accept and act on (e.g. Pittock 2009; IPCC 2014). The overwhelming scientific consensus is that climate change *is* happening and humans are responsible for the increased global temperatures over the last century (Cook *et al.* 2013; IPCC 2014). But so what? Does it matter? Our civilisation evolved in the last 8,000 years in a period of stable climate, but now we are changing that radically. We face a major threat from climate change to the survival of our planet's ecosystems, and our civilisation relies on these. We face major impacts on our food supply due to decreasing water supplies, increased fire frequency, ecosystem change and expanding deserts (Solomon *et al.* 2009; IPCC 2014). We face social dislocation due to displacement of potentially hundreds of millions of climate refugees this century due to sea level rise (Dasgupta *et al.* 2007). So yes, it *does matter*. Climate change action is rightly seen as a key part of reaching sustainability. It is not the *only* thing we need to do, but it is something we must urgently stop denying, so we carry out the solutions (which we know).

Sadly, there is much focus lately just on 'adaptation', as 'mitigation' is proving so hard politically. However, it is not a case of either/or, we have to do *both*, for they are necessary and complementary (Pittock 2009). We do indeed have to adapt, but even more importantly we have to *mitigate*. That means we need a carbon price, we also need action to promote energy efficiency and renewable energy (see next section). It means we need to keep most of our remaining fossil fuels 'in the ground' (Princen *et al.* 2013). However, it will have to go deeper than this, we must also deal with overpopulation and overconsumption (Chapter 7), and the endless growth economy (Chapter 4). Climate change is really a symptom of an *unsustainable culture*. We need to solve its root causes.

> **What you can do!**
>
> Demand urgent action to *mitigate* climate change (not just on adaptation). That means we need to demand a carbon price (Emission Trading Scheme or Carbon Tax), Feed-in Tariffs for renewable energy, and Mandatory Renewable Energy Targets (MRETs). Push *all* levels of government to reduce their carbon footprint, support renewable energy, increase energy efficiency, support 'Factor Five', support sustainable building and efficient and affordable public transport. Reduce your own carbon footprint, while at the same time pushing your local governments and businesses to reduce theirs. Join a climate action group, write letters, go and see your local political representative, and keep pushing your politicians to *act* right now.

6 Appropriate technology: a renewable future

Techno-centrism and Cornucopianism have been the hallmarks of modernism, the idea that technology can 'fix everything'. The environmental crisis shows clearly that it cannot, but technology can actually help – if it's *appropriate* (see Chapter 10). One of the most frustrating aspects is the denial of real and workable solutions using appropriate technology such as renewable energy. We could move practically and economically to renewable energy within two to three decades. So why do we deny this? Chapter 10 explains why, and also lists the various renewable technologies. Civilisation *can* reach a 95 per cent sustainably-sourced energy supply by 2050. However, energy efficiency and conservation must also play a key role in any sustainable energy future (WWF 2011; Diesendorf 2014). In addition, our energy demand cannot keep on growing. If we halve our energy use through energy conservation, then it is feasible to supply this through renewable energy. For years we buried our heads in the sand and denied the potential of renewable energy. However, the technologies exist, are feasible and affordable – all we need is the *political will*. Chapter 10 also discusses three 'inappropriate' technologies: nuclear power, geo-engineering and CCS. These are false solutions, part of denial, and actually hinder us reaching a sustainable future.

> **What you can do!**
>
> Demand 100 per cent renewable energy from your governments *within 20 years*, with immediate strong Mandatory Renewable Energy Targets and 'Feed-in Tariffs' for all types of renewable energy. Oppose construction of further fossil fuel power plants. Oppose the dangerous mirage of nuclear energy. Install renewable energy yourself (e.g. solar hot water, PV). Get a sustainability audit of your home so you can cut energy and water use.

7 Reducing poverty and inequality

As Rees (2010) notes, the world is in thrall to a grand socially constructed vision of global development and poverty alleviation centred on unlimited economic expansion fuelled by open markets and more liberalised trade. However, rather than ending poverty, modernism has led to the environmental crisis, which is now impacting strongly on the poor (who rely on free ecosystem services). The 'trickle down' approach has proven to be cynical spin. Inequality in income distribution has increased within OECD counties, and also in major emerging economies such as China and India (OECD 2011). However, poverty exacerbates environmental problems, which in turn exacerbate poverty. An increasing population in poverty means they push ecosystem services beyond their sustainable level. Less available ecosystem services means the poor have less with which to live. We are in danger of going into a downward reinforcing spiral (MEA 2005).

Brown (2006, 2011) notes that in an increasingly integrated world, eradicating poverty and stabilising population have become 'national security issues'. Slowing population growth helps eradicate poverty (and its distressing symptoms), and conversely, eradicating poverty helps slow population growth (Brown 2006). With time running out, the urgency of doing both simultaneously is clear. One way of narrowing the gap between rich and poor is by ensuring 'universal education', and universal primary education by 2015 is one of the Millennium Development Goals.

The World Bank estimates that $12 billion a year could achieve this for 80 of the world's poorest countries. An important incentive is to get children into 'school lunch programmes'. Girls especially benefit, as they stay in school longer, marry later, and have fewer children (thus easing population pressures). Carrying this out in the 44 lowest income countries would cost an estimated extra $6 billion a year (Brown 2006). A key way of reducing poverty (and consumer pressure) is *greater equality of income* (Wilkinson and Pickett 2010). One important (possibly essential) solution is both minimum and maximum incomes (Dietz and O'Neill 2013).

If the developing world uses old carbon-polluting and resource-wasteful technology to grow its economy, the environmental crisis will markedly worsen. These countries thus need assistance to develop in smarter, *more sustainable* ways. Rich nations should spend 1–2 per cent of their GDP on ecologically sustainable environmental aid to the developing world (in all our interests). After all, we live on the same planet in peril. The developed world also needs to forgive much of the old debt that the developing world still owes (often due to poor projects or corruption). Additional funding to reach basic social goals (so as to break out of the poverty trap) would cost $68 billion a year (Brown 2006). There is also a growing movement towards 'fair trade' with the developing world, which for too long has been forced to sell their products too cheaply.

> **What you can do!**
>
> Support fair trade products. Support 1–2 per cent of your nation's GDP going in ecologically sustainable environmental aid to the developing world. Support the forgiving of much of the old debt owed by developing nations. Argue for greater equality of income, and for minimum and maximum incomes in society.

8 Education and communication

The environmental crisis also represents a failure to educate and communicate. George Perkin Marsh wrote 'Man and Nature' in 1864, and almost certainly had a more accurate view of humanity's role in the natural world than most decision-makers today (Daily 1997). Some 150 years later, most 'educated' people still remain unaware of his basic message: humans are part of Nature and reliant upon it. This fact is a condemnation of schools, universities and the media. It highlights the failure of professional scientists to adequately communicate this (Daily 1997). There is also the problem that education ideology can become an indoctrination that reflects the dominant cultural ideology. For example, Kopnina and Blewitt (2015) note that some critics have branded UNESCO's approach to environmental education as complicit in perpetuating the new 'holy grail' of the dominant political elites – namely, the expansion of consumerist culture. Such an approach will not solve an environmental crisis *created* by an endless growth ideology.

We desperately need to close the 'culture gap' in environmental understanding (Ehrlich and Ehrlich 2010). A key problem is that school curricula almost never cover ecology in the detail needed to explain human ecological dependence on Nature. The problem extends even to universities, which Ehrlich and Ornstein (2010) observe are full of conservatism, entrenched interests and 'frequent sloppy thought'. Chapters 5 and 8 have shown that academia is liable to get bogged down in unhelpful theory, which does not in fact advance sustainability (and may hinder it). The issue is not just education, it is about 'empowerment' to take action. 'Facts' about environmental degradation abound, the trouble is finding how you *empower people* to take meaningful action. The best way to do this remains a 'work in progress'. It is important to aim for 'small meaningful local actions that individuals or small groups can guarantee to carry through to completion' (Hill 1998). However, we desperately need concerted mass collective action as well, as we discuss under 'politics'.

In the United States a number of schools and universities are working to embed sustainability into cultures, including integrating environment science, media literacy and critical thinking into curricula. In Europe, 39,500 schools were awarded a 'Green Flag' for greening their curricula. Some schools are also modelling a sustainable way of living (Assadourian 2013, p. 120). 'Big History' has been found to be a useful tool to educate all walks of life. It builds on the work of 'The Universe Story' (Swimme and Berry 1992). It starts with the Big Bang, then explains the creation of stars, dispersal of elements formed within stars, the story of planets and our solar system, then the history of Earth and the evolution of life. Anecdotal evidence suggests students studying Big History have been able to 'change their reality map', resulting in more sustainable behaviour (Collins *et al.* 2013, p. 222).

One interesting practical way to support sustainability has been the idea of creating 'Life Centres' (Boyden 2004). They would be new public institutions that focus on the processes of life, and the well-being of people and Nature. Networks of Life Centres would focus on the idea of 'healthy people on a healthy planet'. They would play a direct educational role, act as discussion forums and repositories of sustainability information. They would fill a serious gap in the institutional structure of society. Finally we would have an *institution designed for sustainability*. They would help to counteract the tokenism that is rife when discussing 'sustainability'.

What you can do!

Talk about sustainability, about the need for *all three strands*. Condemn tokenism, and argue for serious action in your community. If you have children at school, speak to the school about discussing the environmental crisis and becoming more sustainable. If you are at university, promote these ideas there. Work to empower your local groups to take action. Support the creation of 'Life Centres' and environmental education programmes in your community, and lobby your council to establish these.

9 The politics of it all!

Many of us shy away from politics, but this is really just a form of denial. It is pure laziness, a way of shirking any responsibility to 'do something' about sustainability. If we all do this then nothing will happen. Disdain for politics is really apathy and 'giving up'. Avoiding politics is thus failing to reach sustainability. Many politicians would like to act on the environmental crisis, but need to know they have community support. That means you need to tell them. You need to be politically active on sustainability, you need to write letters to papers and local representatives. Your effectiveness and enjoyment of activism will be enhanced by joining an environment or climate action group (Diesendorf 2009). We know change for sustainability is unlikely to come from the business community (though some have seen the light). Change is unlikely to come from the education system (but hopefully this will change). It is also unlikely to come from governments – *unless we make it happen*. We need to insist our leaders begin a national sustainability planning process and draft international accords needed to implement action for a truly economically-secure, ecologically-stable and socially-just future (Moore and Rees 2013, p. 50). Politics is at a crossroads, the choices are either affluence with persistent disparity, possible collapse, or moderation with prospects for equity. If there is to be some kind of prosperity for all world citizens, the European-US model needs to be superseded, making room for ways of living and consuming that leave a light footprint (Sachs 2013, p. 125).

So don't switch off and ignore politics. As Edmund Burke noted: 'All that is needed for evil to triumph is for good men to do nothing'. The only way to reach sustainability is *activism*. That means political action and political lobbying. The maxim for politicians is 'one letter equals a hundred votes'. So if they get 1,000 letters, it seriously worries them. Hence your contribution *really does count*. Solutions exist, we can reach sustainability, it is not impossible or hopeless. However, one of our major problems is lack of political will. It is no good just sitting back and blaming this on 'politicians'. We elect them, they are our representatives. If they are not taking action, it is because we are not telling them to. We need to accept the political nature of the task of reaching sustainability. Without mass political action, we will not reach sustainability. If we want to reach a sustainable future, then we need to be active in environmental politics *now*. I am not advocating a vote for any particular political party. Whatever party you vote for will need people who tell them about the environmental crisis and the need for sustainability. This is a key part of the 'Great Work'.

What you can do!

Be politically active (in whatever party) on sustainability. Join an environment or sustainability group. Write a letter, or even better, *go and see* your local member. Demand he/she supports change for all the three strands of sustainability. Especially demand your representative supports strong climate change and renewable energy action right now!

The 'Great Work'

So modernism and consumerism have failed to bring a better world, in fact they have caused the environmental crisis. Such a worldview has broken our society, our economy, and is well on the way to breaking the ecosystems on which civilisation depends. Most of us deny this, but if we could overcome this denial, then we could start *fixing things*. Berry (1999) argues that contemporary history has shown three things: 1) The devastation of the Earth; 2) The incompetence of religion and cultural traditions to deal with this devastation; 3) The rise of a new ecological vision of the Universe. We now need an ecocentric ethic or 'Earth ethics' (Rolston 2012), an ethic not based just on enlightened self-interest, but one that considers the rest of Nature too (Assadourian 2010).

Berry (1999) argues our task is to undertake 'The Great Work', and provides an ecocentric vision for this work of Earth repair. He notes that every culture produces a 'Great Work', an overarching vision that 'gives shape and meaning to life by relating the human venture to the larger destinies of the Universe'. Macy (2012) seeks a similar thing in her 'Great Turning', while Rees (2010) calls this 'Survival 2100'. This is a very necessary philosophical and ethical 'grand narrative'. It is an over-arching worldview, a narrative that provides meaning at a time when we desperately need meaning to foster the deep beliefs that will support us in reaching sustainability. It is a dream that people can believe in, a dream that can inspire young and old. It is a dream that can lead to a sustainable future. Martin Luther King inspired people through his statement 'I have a dream!'. He would not have inspired them if he had said 'I have a catastrophe!'. *People need hope*, they need a dream to work towards. Hope is linked to imagination and this allows us to conceive the world differently and create a sustainable future (Collins 2010). Change comes from necessity, hope, realisable aspirations and joy, not shame and blame (Wackernagel and Rees 1996). Given the urgency involved in our predicament, this grand narrative, this 'Great Work' of repairing the Earth is precisely the dream we need, precisely what we need to move forward to an ecologically sustainable future. The 'Great Work' is our path to sustainability, and we won't get there unless we move past denial and take that path. There will be many parts to the 'Great Work' (many are listed here), and they will change over time. However, the big shift is accepting this new worldview of sustainability: *the Great Work of Earth Repair*.

Can we demystify 'sustainability'?

Yes we can, we can demystify 'sustainability'. The key step is to accept reality, accept the gravity of our predicament, roll back denial, and rapidly put in place the solution frameworks covered above. 'Sustainability' is actually saving our civilisation through solving the environmental crisis, and it cannot be a spectator sport (Brown 2006). It is time for us together to change the world for the better. Sustainability is fixing the broken things in our culture. It is *healing the planet*. An unsustainable worldview, a broken economy and broken society, and a breaking Nature. Our current worldview, economics and actions are not rooted in the Earth, but in an idea that we can be 'masters' of Nature. Yet this is an ecological absurdity, one that has caused great grief to the world, and also to the human poor. It is time to wake up, to accept that Nature has intrinsic value, to accept we have come close to a disaster that would impoverish the Earth and future generations. In fact, this disaster is underway. Our roots have always been in the Earth. The 'old' sustainability knew this, indigenous cultures knew (and still know) this. It is not something to be ashamed of or deny. It is something to celebrate and value.

The environmental crisis is telling us we need to change the way we view the world, we need to move beyond modernism (and most postmodernism) and consumerism and move to an ecocentric worldview (Assadourian 2010). It is telling us that our economy is unsustainable, our society is unhappy and anxious, and the living world that supports us is starting to collapse. It is a major, inconvenient and unpalatable truth for a modernist and consumerist society. It is a major reality check. On the other hand, I agree with Hulme (2009) and Pittock (2009) that climate change (and the environmental crisis overall) really is an 'opportunity' to get things right, to become sustainable. The scarcest of all resources is *time*. We cannot just reset the clock, for Nature is the time-keeper (Brown 2006). If someone erects a tombstone for our civilisation, it cannot say we did not understand, for we do. It cannot say we did not have the resources, for we do. It could only say we were too slow to act, and that time ran out (Brown 2006). Action for sustainability is long overdue. The path is there, we do in fact know what to do. *Will we?*

There is no way that business-as-usual is sustainable. 'Sustainability' (to be meaningful) cannot be based on endless growth, hence why it is distinguished here from the 'sustainable development' of 'Our Common Future' (WCED 1987) – which was based on growth. Closing our minds in denial will not change the reality we face. It won't help, in fact it will degrade our well-being. Brown (2006) asks whether the future world will be a world of decline and collapse, or a world of environmental restoration. Can the world mobilise quickly enough? Where will the wake-up calls come from? What form will they take? Will we hear them? We must remember we still have an opportunity to 'get it right', to heal the damage our society, our numbers, our growth economy, our consumerism, our inappropriate technologies, and our carbon fuels have done to the world. 'Sustainability' is the chance to abandon denial and accept reality, ethics, and responsibility. Sustainability is solving the environmental crisis, and we can no longer say it's not 'economic' to do this and protect Nature. We have to, and thus we can make it economic. Sustainability requires it. It certainly won't cost more than the $1,531 billion a year we spend on the military (SIPRI 2010). In fact, Brown (2011) estimates it will cost $185 billion a year, or just 12 per cent of the world's yearly military budget.

If we have the vision, the dream of sustainability based on 'Earth repair', the faith and hope that we can solve the environmental crisis, then 'yes we can'. If we believe we can make a difference then we will. If we despair, give up our hope of solving these issues, if we abdicate our choice to act, then we will fail. We either choose to act for an ecologically sustainable future, or we contribute to a growing environmental disaster (Pittock 2009). However, the situation is not all about doom and gloom, it's about a new future, new appropriate technologies, new government policies, new (responsible and regulated) markets, new opportunities, and a new (or perhaps rediscovered) worldview of how we should live on Earth. Sustainability can be an exciting challenge. It won't be simple or easy, but if we can face and conquer our denial, then our future is exciting as we make a truly sustainable and better world. Can we solve the environmental crisis and reach sustainability? *We can if we choose to.*

Our first hope must be that we prevent collapse by following a new set of philosophical, ethical and cultural norms that bring about a 'life-sustaining' civilisation (Assadourian 2013, p. 302). The solutions are there. But if we are realists, we should also prepare a 'secondary hope', that failing to prevent the 'Great Unravelling', we preserve enough knowledge and wisdom so that as the dust settles, with a stable lower population and consumption, our descendants don't reinvent our mistakes (ibid., p. 303). The times we live in thus seem orchestrated to bring forth from us the greatest moral strength, courage and creativity. We each must seek to rise to this task. This is true *moral integrity*, the moral compass by which we can steer. This is honour to the world that bore us. When things are this unstable, a person's determination, how they invest their energy, heart and mind, can have a much greater effect on the larger picture than we are accustomed to think (ibid.). We desperately need to remember anthropologist Margaret Meade's comment (used with permission, InterculturalStudies 2010): 'Never doubt that a small group of thoughtful, committed citizens can change the world. ® Indeed, it is the only thing that ever has'.

If enough of us undertake the 'Great Work', we will succeed. However, no matter the number of us that take up the task, whether it be millions or billions, we will *make things better* than they would have been if we had stood by and done nothing. The 'Great Work' lies before us. What task could be more inspiring or challenging?

To conclude, we still live in an amazingly beautiful world. It is truly a privilege to be here to witness it. I write this final section on my bit of land on the edge of the half a million ha Wollemi National Park in NSW, Australia. I write from a stone house I built myself from local broken rock and recycled timber. My computer and everything else is powered by the Sun. Outside, a red-neck wallaby grazes in the shade under gum trees. An hour ago, a 1.5 m long goanna (lizard) walked trustingly right past me (and my dog!) as I sat reading in my doorway. Nearby stretches the largest wilderness in the state of New South Wales, an area I fought to make a national park in the 1970s, and to which I will always feel the responsibility of being a custodian. What area could be more sustainable, more in harmony? This provides my personal hope. We *can* demystify sustainability and act on real solutions. We can live in harmony with Nature into the future, we can solve the environmental crisis. And we *should.*

We should feel wonder at (and respect) the Nature of which we are a part. We should care about equity and equality, about human and environmental ethics, about justice at all levels, human justice and eco-justice. We should care about organising an economy that works within the Earth's limits, and that is equitable to all. To be sustainable, we should have a loving engagement and commitment to the world that nurtured us, and still supports us. We should heal the broken things in our culture. *This is what sustainability should be.* It makes good sense on many levels, it is pragmatic, scientific, and ethical. Sustainability requires we encourage the 'best in humanity', that we rise up past our current broken modernist culture, that we become 'fully human' through caring and sharing. Sustainability is within our grasp, we can live in loving harmony into the future with this wondrous world to which we are privileged to belong. Now, to be meaningfully sustainable it is time to demystify sustainability, cut through the waffle and tokenism, and accept both reality and responsibility. It is time to listen to the Earth, foster our compassion, grasp our moral integrity – and *act*.

References

Assadourian, E. (2010) 'The rise and fall of consumer cultures', in *2010 State of the World: Transforming Cultures from Consumerism to Sustainability*, eds. L. Starke and L. Mastny, London: Earthscan.

Assadourian, E. (2013) 'Re-engineering cultures to create a sustainable civilization', in *State of the World 2013: Is Sustainability Still Possible?*, ed. L. Starke, Washington: Island Press.

Atlee, T. (2003) *The Tao of Democracy: Using Co-Intelligence to Create a World that Works for All*, Manitoba, Canada: The Writers Collective.

Berry, T. (1999) *The Great Work*, New York: Belltower.

Boyden, S. (2004) *The Biology of Civilisation: Understanding Human Culture as a Force in Nature*, Sydney: UNSW Press.

Brown, L. (2006) *Plan B 2.0: Rescuing a Planet under Stress and a Civilization in Trouble*, New York: W. W. Norton and Company.

Brown, L. (2011) *World on the Edge: How to Prevent Environmental and Economic Collapse*, New York: W. W. Norton and Co.

Catton, W. (1982) *Overshoot: the Ecological Basis of Revolutionary Change*, Chicago: University of Illinois Press.

Chawla, L. and Cushing, D. (2007) 'Education for strategic environmental behaviour', *Environmental Education Research*, vol. 13, no 2, pp. 437–452.

Collins, D., Genet, R. and Christian, D. (2013) 'Crafting a new narrative to support sustainability', in *State of the World 2013: Is Sustainability Still Possible?*, ed. L. Starke, Washington: Island Press.

Collins, P. (2010) *Judgment Day: The Struggle for Life on Earth*, Sydney: UNSW Press.

Cook, J., Nuccitelli, D., Green, S., Richardson, M., Winkler, B., Painting, R., Way, R., Jacobs, P. and Skuce, A. (2013) 'Quantifying the consensus on anthropogenic global warming in the scientific literature', *Environmental Research Letters*, vol. 8, pp. 1–7, online, available at: www.skepticalscience.com/docs/Cook_2013_consensus.pdf (accessed 22 July 2014).

Daily, G. (1997) *Natures Services: Societal Dependence on Natural Ecosystems*, Washington: Island Press.

Daily, G. and Ellison, K. (2002) *The New Economy of Nature: the Quest to Make Conservation Profitable*, Washington: Island Press.

Dasgupta, S., Laplante, B., Meisner, C., Wheeler, D. and Yan, J. (2007) *The Impact of Sea Level Rise on Developing Countries: a Comparative Analysis*, World Bank, online, available at: http://elibrary.worldbank.org/doi/book/10.1596/1813–9450–4136 (accessed 2 December 2014).

Diesendorf, M. (2009) *Climate Action: a Campaign Manual for Greenhouse Solutions*, Sydney: UNSW Press.

Diesendorf, M. (2014) *Sustainable Energy Solutions for Climate Change*, London: Earthscan (Routledge).

Dietz, R. and O'Neill, D. (2013) *Enough is Enough: Building a Sustainable Economy is a World of Finite Resources*, San Francisco: Berrett-Koehler Publishers.

Edwards, A. (2005) *The Sustainability Revolution: Portrait of a Paradigm Shift*, Canada: New Society Publishers.

Edwards, A. (2010) *Thriving Beyond Sustainability: Pathways to a Resilient Society*, Canada: New Society Publishers.

Ehrenfeld, D. (1978) *The Arrogance of Humanism*, New York: Oxford University Press.

Ehrlich, P. and Ornstein, R. (2010) *Humanity on a Tightrope: Thoughts on Empathy, Family and Big Changes for a Viable Future*, New York: Rowman and Littlefield.

Ehrlich, P. and Ehrlich, A. (1991) *Healing the Planet: Strategies for Resolving the Environmental Crisis*, New York: Addison-Wesley Publishing Company.

Ehrlich, P. and Ehrlich, A. (2010) 'The culture gap and its needed closures', *International Journal of Environmental Studies*, vol. 67, no. 4, pp. 481–492.

Ehrlich, P. and Ehrlich, A. (2013) 'Can a collapse of global civilization be avoided?', *Proceedings of the Royal Society, B*, vol. 280 (20122845 ejournal).

Engelman, R. (2012) 'Nine population strategies to stop short of 9 billion', in *State of the World 2012: Moving Toward Sustainable Prosperity*, ed. L. Starke, Washington: Island Press.

Engelman, R. (2013) 'Beyond sustainababble', in *State of the World 2013: Is Sustainability Still Possible?*, ed. L. Starke, Washington: Island Press.

Gare, A. (1995) *Postmodernism and the Environmental Crisis*, London and New York: Routledge.

Gough, S. and Scott, W. (2007) *Higher Education and Sustainable Development: Paradox and Possibility*, Abingdon: Routledge.

Greer, J. M. (2008) *The Long Descent: a User's Guide to the End of the Industrial Age*, Canada: New Society Publishers.

Greer, J. M. (2009) *The Ecotechnic Future: Envisioning a Post-peak World*, Canada: New Society Publishers.

Haswell-Elkins, M. and Washington, H. (2014) 'Not so cheap: Australia needs to acknowledge the real cost of coal', *Conversation*, 9 June, online, available at: https://theconversation.com/not-so-cheap-australia-needs-to-acknowledge-the-real-cost-of-coal-26640 (accessed 22 July 2014).

Hill, S. B. (1991) 'Ecological and psychological prerequisites for the establishment of sustainable agricultural communities', in *Agroecosystems and Ecological Settlements*, eds L. Salomonsson, E. Nilsson and T. Jones, Uppsala, Swedish University of Agricultural Sciences, pp. 28–51.

Hill, S. B. (1992) 'Ethics, sustainability and healing', Talk to Alberta Round Table on the Environment and Economy and the Alberta Environmental Network, 11 June, online, available at: www.beliefinstitute.com/article/ethics-sustainability-and-healing (accessed 23 June 2014).

Hill, S. B. (1998) 'Redesigning agroecosystems for environmental sustainability: a deep systems approach', *Systems Research and Behavioural Science*, vol. 15, pp. 391–402.

Hill, S. B. (2001) 'Working with processes of change, particularly psychological processes, when implementing sustainable agriculture', in *The Best of . . . Exploring Sustainable Altern-*

atives: an Introduction to Sustainable Agriculture, ed. H. Haidn, Saskatoon, SK: Canadian Centre for Sustainable Agriculture. pp. 125–134.

Hill, S. B. (2014) 'Considerations for enabling the ecological redesign of organic and conventional agriculture: A social ecology and psychosocial perspective', in *Organic Farming, Prototype for Sustainable Agricultures*, eds. S. Bellon and S. Penvern, Dortrecht: Springer.

Hulme, M. (2009) *Why We Disagree About Climate Change: Understanding Controversy, Inaction and Opportunity*, Cambridge: Cambridge University Press.

InterculturalStudies (2010) Margaret Mead quote, online, available at: www.interculturalstudies.org/Mead/biography.html (quote used with permission from Sevanne Kassarjian).

IPCC (2014) 'Summary for policymakers', Fifth Assessment Report, International Panel on Climate Change, online, available at: http://ipcc-wg2.gov/AR5/images/uploads/WG2AR5_SPM_FINAL.pdf (accessed 16 June 2014).

Jickling, B. (2009) 'Environmental education research: to what ends?', *Environmental Education Research*, vol. 15, no. 2, pp. 209–216.

Kopnina, H. (2012) 'Education for sustainable development (ESD): The turn away from "environment" in environmental education?', *Environmental Education Research*, vol. 18, no. 5, pp. 699–717.

Kopnina, H. and Blewitt, J. (2015) *Sustainable Business: Key Issues*, London: Routledge.

Leonard, A. (2013) 'Moving from individual change to societal change', in *State of the World 2013: Is Sustainability Still Possible?*, ed. L. Starke, Washington: Island Press.

Louv, R. (2005) *Last Child in the Woods: Saving our Children from Nature-Deficit Disorder*, London: Atlantic Books.

Macy, J. (2012) 'A wild love for the world', Joanna Macy interview by Krista Tippett, 'On Being', American Public Media, 1 November, online, available at: www.onbeing.org/program/wild-love-world/61 (accessed 11 September 2013).

Marsh, G. P. (1864) *Man and Nature; or, Physical Geography as Modified by Human Action*, New York: Scribner.

MEA (2005) *Living Beyond Our Means: Natural Assets and Human Wellbeing, Statement from the Board, Millennium Ecosystem Assessment*, United Nations Environment Programme, online, available at: www.millenniumassessment.org.

Meadows, D., Randers, J. and Meadows D. (2004) *The Limits to Growth: the 30-year Update*, Vermont: Chelsea Green.

Moore, J. and Rees, W. (2013) 'Getting to one-planet living', in *State of the World 2013: Is Sustainability Still Possible?*, ed. L. Starke, Washington: Island Press.

Moore, K. and Nelson, M. (2013) 'Moving toward a global moral consensus on environmental action', in *State of the World 2013: Is Sustainability Still Possible?*, ed. L. Starke, Washington: Island Press.

NEF (2012) 'Five ways to well-being: postcards', online, available at: www.neweconomics.org/publications/five-ways-well-being-postcards (accessed 22 July 2014).

OECD (2011) *Divided We Stand: Why Inequality Keeps Rising*, Paris: Organisation for Economic Development and Cooperation, online, available at: www.oecd.org/document/51/0,3746,en_2649_33933_49147827_1_1_1_1,00.html (accessed 22 July 2014).

Orr, D. (2013) 'Governance in the long emergency', in *State of the World 2013: Is Sustainability Still Possible?*, ed. L. Starke, Washington: Island Press.

Pittock, A. B. (2009) *Climate Change: the Science, Impacts and Solutions*, Collingwood, Vic/London: CSIRO Publishing/Earthscan.

Princen, T., Manno, J. and Martin, P. (2013) 'Keep them in the ground: ending the fossil fuel era', in *State of the World 2013: Is Sustainability Still Possible?*, ed. L. Starke, Washington: Island Press.

Rees, W. (2010) 'What's blocking sustainability? Human nature, cognition and denial,

Sustainability: Science, Practice and Policy, vol. 6, no. 2 (ejournal), online, available at: http://sspp.proquest.com/archives/vol.6iss2/1001-012.rees.html (accessed 21 June 2014).

Rickinson, M. (2003) 'Reviewing research evidence in environmental education: Some methodological reflections and challenges', *Environmental Education Research*, vol. 9, no. 2, pp. 257–271.

Robinson, K. S. (2013) 'Is it too late?', in *State of the World 2013: Is Sustainability Still Possible?*, ed. L. Starke, Washington: Island Press.

Rolston III, H. (2012) *A New Environmental Ethics: the Next Millennium of Life on Earth*, London: Routledge.

Sachs, W. (2013) 'Development and decline', in *State of the World 2013: Is Sustainability Still Possible?*, ed. L. Starke, Washington: Island Press.

Schmidheiny, S. (1992) *Changing Course*, Cambridge, MA: MIT Press.

SIPRI (2010) 'Yearbook 2010', Stockholm International Peace Research Institute, online, available at: www.sipri.org.

Solomon, S., Plattner, G.-K., Knutti, R. and Friedlingstein, P. (2009) 'Irreversible climate change due to carbon dioxide emissions', *Proceedings of the National Academy of Sciences*, vol. 106, no. 6, pp. 1704–1709.

Solow, R. (1993) 'Sustainability: an economist's perspective' in *Economics of the Environment: Selected Readings*, eds. R. Dorfman and N. Dorfman, New York: Norton.

Soskolne, C. (2008) 'Preface' to *Sustaining Life on Earth: Environmental and Human Health through Global Governance*, ed. C. Soskolne, New York: Lexington Books.

Soskolne C., Kotze, L., Mackey, B. and Rees, W. (2008) 'Conclusions: challenging our individual and collective thinking about sustainability', in *Sustaining Life on Earth: Environmental and Human Health through Global Governance*, ed. C. Soskolne, New York: Lexington Books.

Stern, N. (2006) *The Economics of Climate Change* (Stern Review), London: Cambridge University Press.

Swimme, B. and Berry, T. (1992) *The Universe Story: from the Primordial Flaming Forth to the Ecozoic Era*, San Francisco: Harper Books.

Vitousek, P., Mooney, H., Lubchenco, J. and Melillo, J. (1997) 'Human domination of Earth's ecosystems', *Science*, vol. 277, pp. 494–499.

Wackernagel, M. and Rees, W. (1996) *Our Ecological Footprint: Reducing Human Impact on the Earth*, Gabriola Island, BC, Canada: New Society Publishers.

Wals, A. (2007) *Social Learning: Towards a Sustainable World*, Wageningen: Wageningen Academic.

Wals, A. (2010) 'Between knowing what is right and knowing that is it wrong to tell others what is right: on relativism, uncertainty and democracy in environmental and sustainability education', *Environmental Education Research*, vol. 16, no. 1, pp. 143–151.

Washington, H. and Cook, J. (2011) *Climate Change Denial: Heads in the Sand*, London: Earthscan.

WCED (1987) *Our Common Future*, World Commission on Environment and Development, London: Oxford University Press.

Wilkinson, R. and Pickett, K. (2010) *The Spirit Level: Why Equality is Better for Everyone*, London: Penguin Books.

WWF (2011) *The Energy Report: 100% Renewable Energy by 2050*, Switzerland: WWF, online, available at: http://wwf.panda.org/what_we_do/footprint/climate_carbon_energy/energy_solutions/renewable_energy/sustainable_energy_report/ (accessed 3 February 2012).

Zencey, E. (2013) 'Energy as master resource', in *State of the World 2013: Is Sustainability Still Possible?*, ed. L. Starke, Washington: Island Press.

INDEX

activism: consumption 121–2; effectiveness 207; empowerment 100, 207; fear of change blocking 170; path to sustainability 100, 207; population 118; practicality of action 104, 207; *see also* solutions

adaptive management: co-option 83–4; denial of limits 83; 'doomsday scenario' 83; 'fallacy of misplaced concreteness' 82; 'figure of 8' 82; 'gardener' approach 84; ideology determining theory 85; neo-environmentalism 84; neoliberal ideology influence on 84; 'new conservation' 84; panarchy 82; trivialising environmental crisis 83, 86; twisting ecological theory 85–6; *see also* resilience

advertising: accountability strategies 122; cause of overconsumption 65, 122; factual information 122; global spending 130; outdoor ban 65; resisting change 130; solutions 128; tax 65

Agenda 21 *see* Earth Summit

anthropocentrism: in academia 80; anthropocentric fallacy 141; versus ecocentrism 14, 95, 141; hubris 160; insidious nature 80, 141; in modernism 12, 141, 144; in neoclassical economics 51; in postmodernism 146; in resourcism 142; in sustainable development 143; *see also* human supremacy

appropriate technology: 'appropriate' to crisis 177; baseload power fallacy 178–9; energy efficiency 180; fossil fuel addiction 178; geo-engineering problems 186; green jobs 178; inappropriate technologies 184; 'intermittent nature' fallacy 179; nuclear power problems 184; policy paralysis 119; solutions 203; *see also* carbon capture and storage; *see also* renewable energy, techno-centrism

Berry, Thomas 14, 42, 96, 101, 105, 138, 140, 151–2, 154, 173, 208

biodiversity: biosphere, sustainable 66, 71, 86–8, 101; causes of loss 72, 79; climate change impact 73; collapse *see* ecosystem collapse; end to birth 72; extinction crisis 72; native biodiversity importance 85; resilience 81; stability 81; wilderness importance 86; *see also* ecosystem services, Nature

Brown, Lester 64, 72, 87–8, 104, 118, 160, 171, 178, 196–7, 202, 204–5, 208–9

capital: natural *see* natural capital; social 100–1; built 39, 53, 201; human 53; interchangeable, argument for 39; manufactured means of production 38; substitution of 39–40, 59

capitalism: inequalities of wealth 50; land as part of capital 186; means of production 38

carbon capture and storage: denial, part of 186; economic viability, problems with 186–7; energy costs 186; geological suitability 186; leaks 187

Carson, Rachel 6, 18, 87, 164
Catton, William 76, 79, 115, 139, 163, 173, 196–7
classical economics: description 44; invisible hand 49; Mill, John Stuart 49, 56–7; Smith, Adam 12, 49, 86, 100
climate change: adaptation versus mitigation 203; biodiversity loss 73; carbon pricing 64; conflict 104; crisis 73, 79, 87, 104; denial industry 123, 163–6, 172; denial *see* denial; ethical dimensions 140; extreme weather 73, 104; food impacts 73; geoengineering 185; growth, cause of 203; ocean acidification 73; overshoot symptom 117; planetary threshold 87; regional 79; scientific consensus 202; sea level rise 203; solutions 202; threat to ecosystems 79; water impacts 203
collapse: denial of problems 163; Diamond, Jared 99, 163; economic 51; ecological footprint 79; 'living planet index' 35, 79; long descent 196; overshoot 19, 117, 173, 197; social cohesion 99–100; of societies 19, 99, 129; *see also* ecosystem collapse
compassion: 'care for all' 37, 153; circle of 136, 153; compassionate retreat 139; Earth Charter 25; expansion needed 24, 102; growth in 4; in human behaviour 105; moral circle 95, 146; religious support 97
conservation: Belgrade Charter 143; conservation biology 85–6; environmentalism, distinction from 7, 17; of energy 180; foundations 7; 'new' conservation 84; old sustainability 17; poverty eradication 62; resources *see* resources; versus substitution 59; sustainability, influence on 15; sustainable development impact 36; wilderness importance 7; World Conservation Strategy 20; *see also* adaptive management, Nature
consumerism: advertising impact 122; alternatives 129; choice editing 122, 128; consumer culture 120–1, 129, 202; corporations 122; Cuba 129; dealing with 'more' 129; driver of unsustainability 119; growth presumption 119; historic resistance 120; insatiable demand 121; overconsumption 61, 88, 114, 119, 125, 130; public consumption of services 121; resistance to change 66; social construct 99, 120;

solutions 119–21, 128, 202; spiritual emptiness 105, 151; thriftiness 120, 130; as way of life 120; *see also* factor 5
consumption *see* consumerism
Cornucopianism 19, 61, 139, 177, 203
Costanza, Robert 38, 51, 62–4, 107, 123
'cradle to cradle' 32, 55–6, 88, 116, 121, 124, 128, 136, 202
cultural landscape 148

dadirri 150, 153
Daly, Herman 9, 34, 36, 38, 40, 49–50, 52, 61, 63–6, 105, 122–3, 130, 134, 177
decoupling: failure to stop growth 58; green economy 62; as wishful thinking 58; *see also* factor 5
deep ecology: intrinsic value 145; platform 145; resistance to 146; in sustainability transformation 146; strongest sustainability 42; sustainable development, overlooked by 143; *see also* Naess, Arne
degrowth: definition 55; need for 55; stability question 55
dematerialisation: Business Council for Sustainable Development 122; Daly's rules 123; decoupling limits 124; depletion quotas 123–4; 'extended producer responsibility' 124; 'four Rs' 125; 'natural step' 125; non-renewable resources 64, 88, 123; planned obsolescence 55, 88, 124, 130; renewable resources 40, 64, 123; throughput control 123; tokenism, excuse for 126, 128; *see also* 'cradle to cradle', factor 5
demystification: abandoning endless growth 2, 4, 22, 35, 130, 160, 209; accepting reality 193; broken society 209; clarity needed 1; crucial challenge 36; denial, dealing with 173, 193; drivers of unsustainability 114; feasibility 208–11; need for 9; reader involvement in 3, 193; solving environmental crisis 195; tokenism 195; what sustainability cannot be 193; what sustainability should be 195; worldview importance 199; *see also* solutions, appropriate technology
denial: barrier to sustainability 160; blame-shifting 168; cognitive dissonance 162; commonness 161; as defence mechanism 162; as delusion 162; de-problemising 168; distraction 168; doubt as tool 164; 'elephant in the room' 162; fear of change 170; gambling 171; greenscamming 165;

history 163; *Homo denialensis* 160; ideological basis 166; implicatory denial 167; industry 164; information deficit model 163, 168; interpretive denial 167; literal denial 167; media responsibility 171; ostrich 162; as pathology 163; psychic numbing 160; psychological types 167; versus scepticism 161; silence in 162; 'Silent Spring' 164; solutions to 173; stupid things 160; think tanks 165; ways we let denial prosper 170
Diesendorf, Mark 34, 51, 64, 140–1, 178, 180–6

Earth Charter: broad scope 24; covenants 102; key document 24; intrinsic value of nature 24; 'meaning' of sustainability 24; visionary nature 24
Earth jurisprudence 88, 108, 141, 194
Earth Summit (1992): Agenda 21, 23; anthropocentrism in 23; criticisms of 23–4; Kyoto Protocol 22; precautionary principle 23; Rio Declaration 22
ecocentrism: backlash against 142; bridging great divide 151; definition 141; Earth ethics 152; intrinsic value of Nature 142, 144; land ethic 14, 151; respect for Nature 152; *see also* deep ecology, worldview
ecocide 72, 140
eco-justice 39, 95, 100, 108, 171, 194, 200, 211
ecological economics 37, 54, 61, 64
ecological footprint 1, 35, 71, 79, 83, 117, 124, 129, 201
ecological ignorance 41
ecological sustainability: balance of Nature 11, 42, 79–80; bankrupting Nature 73; 'do we have a problem?' 72; energy flows 74; eutrophication 77, 79; extinction *see* biodiversity; food webs 74; human dependence on Nature 73; ideology; focus needed 80, 85, 89; keystone species 76; meaning of 86; nitrogen cycle 77; NPP, human use 75; nutrient cycles 76; sustainable biosphere 86–7; theory problems 80; toxification 87; *see also* adaptive management; *see also* ecological footprint, ecological theory, ecosystem collapse, ecosystem services
ecological theory: balance of nature 80–1; ideology determining theory 85; imperial versus arcadian ecology 12; twisting of 85; *see also* adaptive management

economic sustainability: advertising tax 122; bank reserve requirements 51; cooperatives 65, 108; corporate law reform 66, 130; finance re-structure 51, 65; green GDP 63, 202; green investment 65; laissez faire economics 166; maximum incomes 65, 108, 202; meaning of 62; non-growth focus 62; resource control 64; subsidy shifting 64; sustainability tariff 66; tax shifting 64; Tobin tax 65; transition towns 63; well-being 61–3; *see also* green economy, degrowth, GPI, steady state economy
ecosystem collapse: ecological footprint 79; examples of 78–9; jargon for 78; keystone species loss 76; 'natural' suggestion 83; non-linear change 78; overshoot 79; paradox suggestion 83; resilience 81; tipping point uncertainty 41, 79; *see also* 'living planet index'
ecosystem energy flows 74
ecosystem services: categories of 77; Daily, Gretchen 72, 76–7, 81, 88, 154; definition 77; degradation of 77–8; tipping points, uncertainty of 41, 79; *see also* Millennium Ecosystem Assessment, TEEB
education and communication: Belgrade Charter 143; clear objectives 192; education for all 118; education for sustainable development 33, 143, 147, 152, 192; education system failure 152, 205; empowerment 200, 206; environmental education 143, 152, 205; in green economy 62; as ideology 143; ignorance of decision-makers 171, 205; as key solution 205; life centres 206; relativism in 143, 147, 192
Ehrlich, Paul and Anne 12, 18–19, 78, 114–15, 118, 154, 163–5, 171, 197–9, 206
empathy 99, 105, 150, 152–4, 200
environmental crisis: ability to solve 193, 195–6, 199; climate *see* climate change; cost of solving 209; denial of 159, 163–4, 193; energy use 180; externality approach 39, 53; growthism 34, 54; ignoring of 75, 79, 86, 95, 115, 123; 'not too late' 196–9; old sustainability 8; politics 207; solutions to 199; solving crisis = sustainability 195; substitution as cause 39, 59–60; summary of 72; technology impact 177; urgency 170, 190, 192; weak sustainability 39;

environmental crisis *continued*
  see also anthropocentrism, consumerism, growth, modernism, population, postmodernism, solutions
environmentalism: co-option of 8, 85, 128; first wave 36; neo-environmentalism 8, 84; relation to conservation 7, 17, 36; second wave 36; significance 196; third wave 37
ethics: change happening 36, 140; cultural evolution 154, 199; divorce from economics 60; Earth 66, 83, 152, 177, 187, 195, 208; Earth in liquidation 61; ecological imperative 140; environmental 41, 95, 108, 141, 151–3, 200, 211; of economics 60; -'isms' 140; moral circle 95, 146; Pachamama movement 140; as personal tastes 60; respect for Nature 14, 152; rights of Nature 88, 95, 108, 141, 144, 197; teleology (purpose) 60; *see also* compassion, Earth jurisprudence, eco-justice, land ethic, worldview
exponential growth 56, 106, 171
extinction *see* biodiversity

Factor 5 58, 64, 88, 124
food webs 74–5, 81
free market: deification of 52, 66, 160; ecosystem services problems 53; failure of 52, 167; fundamentalism 166; ideology 35; invisible hand 49–50, 52, 160; liberty 166; market-centric model 62; private market value versus public benefits 123; regulation of 18, 37, 50–1, 88, 101, 128, 164, 166–7, 170; *see also* neoliberalism
future generations, knowing what they want 191

geodiversity 145, 150
geo-engineering: carbon dioxide removal 186; in denial 186; problems 186; solar radiation management 185; sulphate aerosols 186
GPI (Genuine Progress Indicator) 63, 65
Great Work: challenge of 4; definition 208; Earth repair 173; ecocentric vision 208; healing broken things 195, 208; hope and imagination 208; overarching vision 208; renewable energy, part of 183; task of sustainability 208; true economic sustainability 66; *see also* Berry, Thomas

green economy: absolute decoupling 62; failing well-being 62; growth assumption 25, 32, 61; low carbon and resource use 62; omission of population 62
governance: administration 102; covenants 24, 102; deliberative civic engagement 103; democracy 101, 103, 108, 142–3; environmental politics 108, 207; fourth strand 103; 'good' 102; 'magic bullet' 103; narcissism 101; 'never get there' syndrome 103; participatory democracy 103, 108; pluralistic governance imbalance 102; purpose of 102; reflexive 103; transition management 104; *see also* Earth Charter
growth: assumption of 54, 119; business as usual 25, 35, 58; cancer cell ideology 56; cause not cure 4, 36; denial around 159–60; versus economic development 57; energy 180; evermoreism 34, 63, 129; green growth 32–3, 80; growth as 'hope' 97; growthism 34, 54, 136; growthmania 57; hypergrowthmania 57; jobs relation to 63; limits to 19–20, 52, 56, 83; myth of endless 33, 56, 138; solution to everything 56; twisting meaning of 56; types of 57; *un*economic 57, 129; *see also* decoupling, solutions

harmony 2, 4, 6–9, 11, 14, 31–2, 42, 80, 85–6, 96, 117, 139, 143, 152, 154, 193, 195, 210–11
human supremacy: anthropocentrism, part of 142, 194; 'common sense' view 142; delusion of 142

ideology: cancer cell 56; death 60; definition 137; denial, ideological basis 166; in education 33; global capitalism 85; growthism *see* growth; industrialism 50, 140; laissez faire economics 166; resourcism 144, 187; in theory 85; *see also* consumerism, free market, mastery of Nature, Nature scepticism, modernism, neoliberalism, postmodernism
imperial versus arcadian ecology 12
intrinsic value of Nature 1, 14, 24, 41–2, 95, 108, 137, 141–2, 144–6, 151–3, 194, 200, 209

justice: belief in 105; delinking from development 99; eco-justice 39, 95, 100, 102; environmental 95; for whom 99;

holistic needed 95, 100; learning to take less 100; modernism, impact of 105; narrow view of 95; peripheral view 105; providing meaning 105; social justice focus 84, 95; *see also* eco-justice

keystone species 75

land ethic 14, 17, 151
Leopold, Aldo 14, 17–18, 74, 143–4, 152
'limits to growth': accuracy of model 19–20; criticism of 19, 56, 83; predictions of collapse 19; thirty year update 19
'living planet index' 35, 79

MacNeill, Jim 1, 29, 35, 104
Macy, Joanna 12, 139, 191, 198, 208
'mastery of Nature' ideology 86, 136, 138, 177, 187, 194
Meadows, Donella and Don 19–20, 33, 56, 72, 83, 88, 107, 122, 137–8, 177, 195–6
media: accuracy, lack of 164, 172; balance as bias 172; bias re climate change 166; bias re population 115; conservative influence 154; controversy 171; denial in 160, 172; disinformation service 164, 171; failure to explain crisis 72; problems 164, 171
Millennium Ecosystem Assessment (MEA) 73, 77–8, 144, 196
modernism: anthropocentrism in 12; anti-ethics 105; dominance of 11; driver of unsustainability 97; globalisation 14, 87; historical movement 11; ideology of 144; impact 11–12, 14, 97, 208; need for change 187, 209; revolt against 12, 140, 144; utilitarian approach 144
Muir, John 13–14, 17–18, 144, 150
myths: consumerism 120, 138; corporate; growth 56, 128, 138; green growth 33; population 114, 118; renewable energy 178–9; re-writing myths 169; shrunken identity 151; trickle-down effect 98

Naess, Arne 14, 139, 141, 145
natural capital: anthropocentric term 39; 'constant natural capital rule' 40; critical 39; definition 38; depletion of 19, 38, 62, 64, 123; externality 39; need to rethink 39; substitution of 19, 40; *see also* sustainability (strong and weak)
natural capitalism 38
Nature: balance of 11, 79–81; being part of 9, 18, 73, 141, 149, 151, 205; custodianship 80, 192; energy flow 74; history of caring 8; human artefact debate 147–8; human dependence on 73; 'law' 6, 9–10; love of the land 11, 150; as machine 12, 14, 142; Nature deficit disorder 150, 191; nutrient cycles 76; resourcism 144, 187; respect for 14, 152; rights of 88, 95, 108, 141, 197; sacredness of 8–11, 14, 80, 139, 150–1; sense of place 149; stewardship 8, 54, 103, 197; wild 12, 83, 146; *see also* Earth jurisprudence, ecosystem services, harmony, intrinsic value of Nature, Living Planet Index, modernism, natural capital, Nature scepticism, sense of wonder, substitution of capital
Nature scepticism: biophobia 149; cooperation versus competition 149; cultural landscapes 148; culture part of Nature 149; 'human artefact' debate 147–8; human/Nature dualism 4, 148; 'influence' versus 'creation' 148–9; questioning reality 146–7
neoclassical economics: assumptions of 51; circular theory of production 52–3; classical economics 49; dominance of 54; entropy 52–3, 55; ethics of 60; externality 39, 53; as fantasy 56, 64; financialisation 51; growthmania 56; *Homo economicus* 9, 105; industrialism 50; invisible hand 49; Keynesian economics 49; limits, refusal to accept 52; modelling 53; reductionism 59; 'resources are infinite' 59; synthesis 49, 54; *see also* capitalism, free market, substitution of capital, thermodynamics
neoliberalism: definition 31; dominance 31, 85, 140; free market as God 31, 105, 127, 195; influence on sustainable development 33, 106; obstacle to sustainability 84, 106; problem for sustainability governance 98, 106
'new' conservation, problem of 84
nuclear power: carbon neutral fallacy 184; in denial 185; dismantling reactors 185; false solution 185; finite high-grade uranium 184; problems 184–5; proliferation of weapons 185; slowness to deploy 184; storage risk 185; waste 185; weak economics 184

old sustainability *see* sustainability
optimism and pessimism 196–9

'Our Common Future': Brundtland, Gro Harlem 21–2; development *plus* environment protection 2, 21–2; 'forgotten imperative' 104; fundamental components 21; growth assumption 22, 31; input/output analysis of 22; Tokyo declaration 22; *see also* WCED

politics: activism needed 100, 207; avoidance of 101, 207; at crossroads 207; environmental 108, 207; key solution 207; lack of political will 101, 207; 'never getting there' syndrome 103; part of Great Work 207; progressive 106; *see also* governance
population: 'anti-human' argument 115; avoidance by the Left 116; 'coercive' fallacy 117; denial of 115; diabolical policy issue 116; ecologically sustainable population 117; exacerbating environmental impacts 117; family planning 116–18, 130; feeding world 117; heresy of 'more' 129; I = PAT 19, 115; increasing affluence impact 116; knee-jerk reaction 115; 'more is better' 160; overpopulation 114, 117–8; passion around issue 115; solutions 118; tragic stalemate 116; world population projections 115
postmodernism: in academia 146, 148; in anthropocentrism 147; definition problem 146; grand narratives, denial of 146; pluralism 147; reality, questioning of 147; 'reason is suspect' 146; relativism 146–7; solutions, inability to provide 146; themes 146; *see also* Nature scepticism
poverty: downward reinforcing spiral 204; environmental degradation 204; fair trade 205; overseas aid 205; reducing 204; trickle-down effect 98, 204; universal education 204

reality: accepting 172, 193; co-creation of 146; denial of 36, 147, 193; ecological 36, 39–42, 48, 56, 62, 80, 86, 139, 160, 194; naive realism 147; questioning of 172; *see also* Nature scepticism, postmodernism
realism 139, 147, 193, 195–6, 198
redesigning ourselves 200
Rees, William xii, 32, 34, 38, 41, 52, 56, 66, 72, 101, 105–6, 120–1, 127, 138, 159–60, 169, 172, 204, 208
renewable energy: ability to supply need 179; availability of 178; 'baseload' versus reliability 178; biochar potential 183; bioenergy 182; denial of 179; energy efficiency 180; energy growth, stopping 183; geothermal power 181, 100% renewables 179; hydropower 181; investment in 179; as key solution 203; lack of political will 183; myths around 179; solar photovoltaic (PV) 181; solar thermal 181; viability; wave and tidal power 182; windpower 180; *see also* appropriate technology
resilience: definition 81; justifying development 84–5; replacing sustainability 84; relation to diversity 81; thinking 84; *see also* adaptive management; *see also* new conservation
resources: Daly's rules 123; 'infinite' claim 59–60, 164; non-renewable 38, 40, 64, 88, 102, 123; renewable 38, 40, 64, 123; substitution of 19, 39–40, 59–60; Zeno's paradox 60; *see also* natural capital, resourcism
resourcism 144, 187
Rio + 20 Summit (2012): 'The Future We Want' statement 25; criticisms of 25–6; failure re climate change 25; *see also* green economy
Rolston III, Holmes 17, 33, 39, 41, 51, 66, 83, 85–7, 95, 100, 108–9, 136, 141, 144–5, 148–9, 152–3, 177, 208

Schumacher, E.F. 38, 56, 96, 129, 177
sense of wonder: becoming fully human 154, 211; belonging 149; bridging great divide 151–2; celebration of Nature 153; creativity 155; dadirri 150, 153; empathy 152–4; imagination 153, 200–1; inspiration 89; listening 150; love of the land 11, 150; meaning, providing 151; rejuvenation of 153; sacredness 8, 14, 150–1; sense of place 149; spiritual intelligence 153; transcendent moments 150; *see also* spiritual values, wilderness, worldview
scepticism: climate change 164–6; definition 161; versus denial 161; *see also* denial, Nature scepticism
'Silent Spring' 6–7, 18, 87, 164
'small and easy' approach *see* triple bottom line
social sustainability: 'baking a bigger cake' 17; components of 94, 107; dealing with change 104; dystopia 95–6; equality of income 96–9; equity 96; Gini coefficient 99; 'human nature' portrayal 99, 105; hunter-gatherer societies 96; meaning

of 107; justice, holistic 100; narcissism 101; practicality of; social capital 100; social cohesion 104; social inclusion 100; social justice 95, 99, 104, 107–8; social problems 98; society, broken 94, 96, 99, 105–6, 195; 'The Spirit Level' 98; technology in 96; Utopia 95; war and conflict 104; wealth redistribution 97; *see also* consumerism, governance, justice
solutions: Band-Aid failure 139; clear objectives 192; climate change 202; conscious evolution 199; cost of 202, 209; dealing with denial 194–5; difficulty involved 195; Earth repair 42, 208–9; future generations' wants 191; hope 196, 198, 207–10; 'is it too late?' 196; long emergency 196, 198; moral integrity 198, 210–1; opportunity 197; perfect storm 198; pluralism versus specificity 192–3; political will 204, 207; poverty and inequality 204; reality, acceptance of 195; redesigning ourselves 200; scale of problem 196; solutions, key 199–207; up to us 198; vision needed 192; 'what sustainability cannot be' 193; 'what sustainability should be' 195; *see also* appropriate technology, consumerism, demystification, education and communication, Great Work, population, postmodernism, renewable energy, steady state economy, worldview and ethics
spiritual values: anti-spirituality 151; creativity 2, 149, 153, 191, 210; empathy 99, 105, 150, 152–3; humility 8, 54, 152; inspiration 89; meaning 151; sacredness 8, 14, 150–1; spiritual emptiness 121; spiritual intelligence 153; *see also* sense of wonder
stability 30, 56, 81–3, 85–6, 100
steady state economy: circular economy 55; dematerialisation 64, 124–5, 128; ecological economics 54; entropy 52–3; 'fallacy of misplaced concreteness' 58–9; Great Depression, not same as 64; jobs in 62–3; key aspects 54; stable population 54; subsidy shifting 64; tax shifting 64; throughput 54, 57–9, 66, 87–8, 122–5; viable path to 64; vision of economy 55; Zeno's paradox 60; *see also* degrowth, ecological economics, economic sustainability
stupid things, believing in 160
substitution of capital 39–40, 59
sustainability: definition, lack of 1–3, 33–5; demystification, why needed 1, 3, 7, 193–5; meaningful 4, 24, 86, 151, 186, 193, 195–6; 'old' sustainability 6; solutions for 191, 199–207; strands of 47, 71, 94; strong 38–41; strongest 42; sustainababble 1, 193; sustainable development, difference to 33–6; transition to 30, 65, 104; weak 38–9; 'what it cannot be' 193; 'what it should be' 195; *see also* ecological sustainability; *see also* economic sustainability, governance, social sustainability, tokenism
sustainable development: anthropocentric bias 143; assumed equivalence with sustainability 33–6; co-option of 8, 37; definition 21; destination or journey 29; development, meaning of 34; failure re conservation 34, 143; growth basis 30–3; masking vested interests 32; meaning of 33–6; oxymoron 2, 29, 33–6, 194; 'sleight of hand' in 32; 'sustainable growth' 30–3; transition strategy 30, 65, 104, 137–9; *see also* 'Our Common Future', tokenism, triple bottom line, WCED
Suzuki, David 8, 11–12, 80

techno-centrism 19, 59, 96, 160, 187, 203
technology: appropriate 96, 177–8, 180, 203; Cornucopianism 19, 61, 139, 177, 203; energy conservation 64, 180, 187, 204; Faustian covenant 61; inappropriate 184–7, 204; mastery of Nature 177, 187; *see also* renewable energy, techno-centrism
TEEB ('The Economics of Ecosystems and Biodiversity') 1, 80–1, 144
thermodynamics, laws of 52–4, 75, 119
Thoreau, Henry David 13–14, 17–18, 144, 154
thresholds/ tipping points 40–1, 52, 79, 87, 154, 169, 199
tokenism 1, 4, 23, 35, 37, 126, 128–30, 193, 195, 206, 211
'tragedy of the commons' 119
triple bottom line (TBL): accounting framework 125; corporate subversion 128; criticism of 126; eco-effectiveness 125–6, 187; eco-efficiency 126–8; ecological modernisation 129; greenwash 128, 194; 'people, planet, profit' 125; rebound effect (Jevons paradox) 127; signs of change 128; 'small and easy' approach 127; as solution 125, 128; Trojan horse for tokenism 126

urgency: denial as obstacle 170, 194; Nature as time-keeper 209; scarcity of time 209; society slow to act 209; wake-up call 209

values: aesthetic 12, 77–8; cultural 11, 77–8, 120; ecocentric 41, 59, 86, 137, 139, 142–4, 147, 187, 208; enoughness 54; failure in 170; monetary 123; non-market 123; shared community 49, 100, 107; wilderness 86, 88, 148–9; *see also* intrinsic value of Nature; *see also* spiritual values

Victor, Peter 35, 48, 54, 56, 58, 62–3, 122

WBGU (German Advisory Council on Global Change) 62, 64, 177–8, 185
WCED (World Commission on Environment and Development) 1, 6, 17, 21, 31, 33–6, 95, 104, 173, 193–4, 209
what you can do 199–207
wilderness: confusion around term 7; conservation significance 7, 86, 88; denial of need for 85, 164; destruction, justification of 85; human artefact debate 147–8; large natural area 7, 86, 88; sense of wonder, restoring 149; 'state of mind' claim 148; value 86, 88, 148–9; wilderness knot 7; *see also* Nature scepticism
wisdom of the elders 8–9, 11, 89
worldview: biounderstanding 139; bridging great divide 151; Catton's five approaches 139; central importance 137–9, 199; compassionate retreat 139; ecosophy 139; Einstein, Albert 136; grand narratives 146–7, 193, 208; growthism *see* growth; humanism 142, 153; instrumental value 14, 141, 172; memes 105, 137; moral compass 210; moral evil 72; Nature as machine 12, 14, 142; paradigm 137; pluralism 143, 147, 192; spiritual capital and intelligence 153, 201; utilitarian approach 41, 84, 141, 144; *see also* anthropocentrism, deep ecology, ecocentrism, ethics, human supremacy, ideology, intrinsic value of Nature, 'mastery of Nature', modernism, Nature scepticism, neoliberalism, postmodernism, realism, resourcism, sense of wonder, solutions, values
Worster, Donald 12, 50, 85